MODELLING FOREST DEVELOPMENT

FORESTRY SCIENCES

Volume 57

Modelling Forest Development

by

KLAUS VON GADOW

Faculty of Forest Sciences and Woodland Ecology,
Göttingen, Germany

and

GANGYING HUI

Academy of Forest Sciences,
Beijing, China

KLUWER ACADEMIC PUBLISHERS

DORDRECHT / BOSTON / LONDON

A C.I.P. Catalogue record for this book is available from the Library of Congress.

ISBN 1-4020-0276-9
Transferred to Digital Print 2001

Published by Kluwer Academic Publishers,
P.O. Box 17, 3300 AA Dordrecht, The Netherlands

Sold and distributed in North, Central and South America
by Kluwer Academic Publishers,
101 Philip Drive, Norwell, MA 02061, U.S.A.

In all other countries, sold and distributed
by Kluwer Academic Publishers,
P.O. Box 322, 3300 AH Dordrecht, The Netherlands

Printed on acid-free paper

Printed in the Netherlands

Contents

Acknowledgements

The compilation of the material included in this book, which gradually developed as a result of our mutual interest in quantitative aspects of forest modelling, was completed during a sabbatical period at the newly established Campus of the University of Santiago de Compostela, the *Escola Politécnica Superior* in Lugo. Lugo is one of the four provincial capitals of Galicia in northern Spain, which is possibly the most productive timber growing area in Europe. The generous hospitality and support of our friends in Lugo, including those of the medical profession, is gratefully acknowledged.

Comments received from Dr. Harold Burkhart are gratefully acknowledged. Dr. Jürgen Nagel of the *Niedersächsische Forstliche Versuchsanstalt* and Matthias Albert and Hendrik Heydecke of the *Institut für Forsteinrichtung und Ertragskunde*, University of Göttingen, helped us develop some of the algorithms and examples. Sonja Rüdiger assisted with the graphs and, together with Jeanne Wirkner and Marga von Gadow, did the major part of the proofreading.

The modelling applications are not limited to a specific geographical region or system of forest management. We have made an attempt to include examples from different countries in Europe, Asia, North America and Africa, covering even-aged plantation forestry as well as uneven mixed forests managed in the selection system. Several of the growth models presented in this text were developed using data sets provided by the *Academy of Forestry in Beijing/China*. We were also able to use stem data collected by the *Niedersächsische Forstliche Versuchsanstalt* in Göttingen for fitting taper equations.

Klaus v. Gadow   Gangying Hui

Chapter One

Introduction

In an unmanaged woodland, forest development follows a succession of periods of undisturbed natural growth, interrupted by intermediate loss or damage of trees caused by fire or wind or other natural hazards. In a managed woodland, the most important periodic disturbances are the thinning operations, which are often carried out at regular intervals and which usually have a significant effect on the future evolution of the resource. Thus, a realistic model of forest development includes both natural growth and thinnings.

The key to successful timber management is a proper understanding of growth processes, and one of the objectives of modelling forest development is to provide the tools that enable foresters to compare alternative silvicultural treatments. Foresters need to be able to anticipate the consequences of a particular thinning operation. In most cases, total timber volume is not a very appropriate measure for quantifying growth or yields, or changes caused by thinning operations. Yield in economic terms is defined by the dimensions and quality attributes of the harvestable logs, and estimating timber products is a central issue of production-oriented growth and yield research.

Growth modelling is also an essential prerequisite for evaluating the consequences of a particular management action on the future development of an important natural resource, such as a woodland ecosystem. Growth models can provide key information about the dynamic change of the less tangible characteristics of a forest, such as the *stability* and *resilience* in an environment affected by industrial pollution or the aesthetic value of a given forest structure.

At a time when single-tree selection systems and uneven-aged forests are the fashion of the day, it is useful to remember that a large proportion of the world's forests are even-aged. Even-aged forests are not necessarily always characterized by straight planting rows and regular spacing. They may include two-storied stands of a shade tolerant species with dominant and suppressed individuals of the same age, or mixed forests with several species of trees, each representing a specific growth rythm and shade tolerance. Even-aged forests include regular populations of trees with very similar attributes where simple modelling approaches are adequate. An even-aged forest may also include trees of different dimensions and different growth patterns, and the resulting structural variation presents a challenge requiring more sophisticated methods.

The most appropriate modelling technique is determined by the level of detail of the available data and by the level of resolution of the required forecast. Thus, a stand model will be used if average population values and area-based information such as the basal area is available. An individual tree model is appropriate when the relevant size attributes of a tree and the attributes and coordinates of its immediate neighbours are available.

Most growth models are developed on the basis of empirical data collected in trials of varying design. Permanent growth series, which are rather common in Central Europe, have been and are being observed and remeasured over long periods of time. Temporary plots, distributed over a wide range of ages and growing sites, are sometimes used to avoid the long wait, which is a disadvantage of the permanent growth series. A compromise may be found by using *interval plots*, which are also spread over a wide range of initial system states, but are measured only twice to include one growth interval.

Types of forest models

Forest models represent average experience of how trees grow and of how forest structures are modified. The level of detail of these models differs greatly. *Tree models* deal with morphological details of branching, stem form and root growth. *Regional production models* and *stand growth models* produce aggregate information about the development of a population of trees with a given set of environmental conditions and given intermittent modifications of stand attributes through human interference and other disturbances.

A forest model enables reasonable predictions to be made about tree growth and stand development. This may be achieved in different ways, depending on the priorities of the user. Timber yield models are only concerned with the commercially relevant aspects of a tree. The ability to simulate the development of a single branch may be merely enlightening or edifying to one user, but vital to another. The ultimate dream of many forest modellers is a comprehensive system with sufficient depth to explain the elementary processes of tree growth, - the commercially relevant yield information would be a welcome by-product.

Most forest models have been developed for specific purposes. They differ with regard to *generality*, - the applicability of the concept to a range of instances, and *precision*, - the degree of exactness in the predictions (Sharpe, 1990). High precision is often achieved at the cost of low generality, and the choice of model usually represents a compromise. A hierarchy of the most common types of forest models is shown in Fig. 1-1. Among the common types of growth models are highly aggregated volume-over-age models for regional yield forecasting, stand models for predicting basal area growth as a function of age and initial basal area, distance-independent size class models for predicting the movement of diameter distributions and distance-dependent individual tree models requiring spatial information in two or three dimensions.

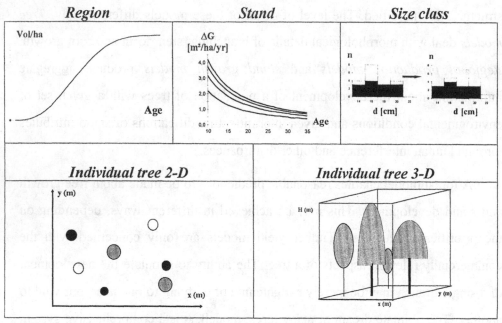

Figure 1-1. Common types of growth models: highly aggregated volume-over-age models for regional yield forecasting, stand models for predicting basal area growth as a function of age and initial basal area, size class models for predicting the movement of diameter distributions and individual tree models requiring spatial information.

Regional yield models are represented by highly aggregated yield-over-age equations. They are used in resource forecasting, specifically for predicting the development of a given age-class distribution in response to a series of periodic harvest levels. *Stand growth models* require more information and, in turn, provide a greater degree of detail, including estimates of dominant height, basal area and stems per hectare. The basic modelling unit in a *size class model* is a representative tree impersonating a number of trees within a size class or cohort. Size class models, requiring even more detailed information than stand models, are probably the most common type for simulating alternative silvicultural programs. *Individual tree models* utilize information about the position and size of specific trees and of the trees in their immediate neighbourhood. The spatial information may be available in two or three dimensions. Three dimensional spatial models are used to quantify the amount of shading and constriction of the growing space, caused by neighbouring trees.

The ability to utilize available data at varying levels of resolution using different techniques within one integrated modelling environment is comparable to using a telescope which can be extended to show varying levels of detail. The practical implementation of this concept is most frequently accomplished in systems which link stand level and size class models (Fig. 1-2).

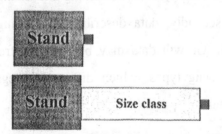

Figure 1-2. Integrated modelling systems are used to cope with varying levels of resolution of the available data, a concept which may be compared with a telescope that can be extended to reveal different levels of detail.

An essential requirement of the telescope approach is *compatibility*, meaning that models of different resolution should produce identical results (Burkhart, 1987). Obviously, the degree of detail of the information that is generated depends on the detail of the input data. However, it is not always easy to link models of

varying resolution in such a way that the system is compatible. One approach is
to assume that the less detailed information on the higher levels of the hierarchy
is more reliable.Thus, a stand level model is considered more reliable than a size-
class model, and compatibility may be achieved by specifying constraints which
are imposed from the higher level of the modelling hierarchy. The high-resolution
model at the lower level is constrained to produce output in compliance with the
output at the higher, low-resolution level. Various techniques are used to attain
compatibility, some of which will be presented in Chapter Five.

Data requirements

A population of trees growing on a given site develops during a succession of
periods of undisturbed growth and intermittent modifications of density and
structure through thinnings. Forest development is a direct response to various
types and intensities of thinnings, and is influenced by the environmental factors
existing on the site. In consequence, two different kinds of empirical data are
required for modelling. Firstly, data describing the change of state variables
through thinning and secondly, data describing the change of state variables
through natural growth. Growth data may be obtained from a variety of field
experiments. The following types, which differ with regard to the dominant
objective, are most common:

- *Provenance trials* for evaluating the suitability of exotic species and/or
  specific provenances on particular sites,
- *Fertilizer trials* for investigating possible growth improvements in
  response to fertilizer application,
- *Spacing and thinning trials*, for evaluating the effects of different
  planting espacements and thinning treatments on tree growth.

These field trials are established and maintained by government research organizations, university departments and company research and development sections. Forestry research and research funding has become increasingly market-oriented in Australia, South Africa and New Zealand, and in many cases, it is very difficult to distinguish between public and private research (Leslie, 1995). Private funding of growth research is mainly found in areas characterized by successful forest industries, e.g., in the Scandinavian countries, in the United States and in many parts of the Southern Hemisphere. Examples of privately-funded growth research conducted by universities are the so-called *Research Cooperatives*, e.g., the *Plantation Management Research Cooperative* at the University of Georgia or the *Loblolly Pine Growth and Yield Research Cooperative* at the Virginia Polytechnic Institute and State University in Blacksburg. Public forestry research funding is a dominant feature in Europe, supporting a number of forestry research institutes. Examples are *INIA* in Spain, *INRA* in France, *METLA* in Finland, the *WSL* in Switzerland and the Federal Forest Research Organizations in Germany. The *Niedersächsische Forstliche Versuchsanstalt* in the federal state of Lower Saxony in Germany, for example, maintains a very large number of permanent growth trials, the oldest plot having been under observation for 113 years (Niedersächsische Forstliche Versuchsanstalt, 1996).

Our present knowledge about the dynamics of different types of forest ecosystems is based mainly on the data obtained in permanent growth trials remeasured on successive occasions, including remeasurements during periods of war (Fig. 1-3).

Figure 1-3. Permanent growth trial for beech (Fagus sylvatica) *at Dillenburg in Germany. The trial was established in 1872; photo taken by R. Schober.*

The limited availability of research funds and the increasing complexity of the questions that are being adressed by research, necessitate a continuous evaluation of the optimum design of growth trials. There is no unique, correct or *normal* silviculture as implied by the normal yield tables, because there is no unanimity about the optimum treatment of a given forest. The manner in which forests are being managed is continually changing. This requires a continuous re-orientation of trial objectives.

Thus, in view of the continuous change of objectives (and environmental conditions) it might not even be necessary or desirable to maintain growth trials for long periods of time, as long as one is able to predict forest development for any arbitrary set of conditions.

Three types of growth trials may be distinguished with regard to the time horizon. *Permanent plots* are established for collecting yield table data for a particular silvicultural program. The plots are remeasured, usually at regular intervals, until harvesting. *Temporary plots*, measured only once, provide age-based information about relevant state variables which is used to construct a yield table, again assuming normal or representative silviculture. *Interval plots* are remeasured once, thus providing an average rate of change in response to a given set of initial conditions. They may be abandoned after one measurement interval.

Permanent plots

The observations from long-term growth plots represent a very important data base for developing growth models. During long periods of time, the change of qualitative and quantitative tree attributes is assessed reiteratively in the same plot. Observations are thus obtained, permitting the construction of a growth model for a given limited set of conditions. Many yield tables were constructed using such a long-term data base.

One of the advantages of a data base derived from permanent plots is the potential to describe polymorphic growth patterns by evaluating the data of each plot separately and by expressing the parameters of the height model as a function of site index or as a function of specific site variables. In this way, it is possible to develop non-disjoint polymorphic height models (Clutter et al., 1983;

Kahn, 1994) and disjoint polymorphic site index equations, which depict the site-specific development of the dominant stand height over age (Fig. 1-4).

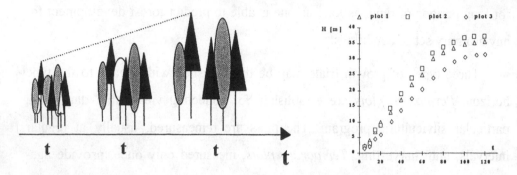

Figure 1-4. Left: A permanent plot with three successive measurements (white trees are removed during thinning operations; the time axis is designated t). Right: Hypothetical data series derived from three permanent plots.

A widely used model for describing the development of a variable over age is the Chapman-Richards equation, which may be expressed in the following form:

$$H = a_0 \cdot \left[1 - e^{-a_1 \cdot t}\right]^{a_2}$$ 1-1

with H   =   dominant stand height [m]
      t   =   stand age [years]
      $a_0 .. a_2$ =   empirical model parameters.

By using permanent plots, it is possible to calculate a separate set of parameters for each plot. The model is *polymorphic* if the parameters determining the shape of the growth curve, a_1 and a_2, are expressed as a function of the site index (the expected height at a given reference age) or as a function of specific site variables, such as the soil depth, the geological formation or the mean annual precipitation. A recent example of a polymorphic height model is presented by Jansen et al. (1996). Many of the existing yield tables are based on permanent plots (Schober, 1987; Jansen et al., 1996; Rojo and Montero, 1996).

A disadvantage of the permanent plot design is the high maintenance cost of the research infrastructure and the long wait for data. The object of the trial is not always achieved, as plots may be destroyed prematurely by wind or fire, or by unauthorized cutting.

Temporary plots

Temporary plots may provide a quick solution in a situation were nothing is known about forest development. They are measured only once, but cover a wide range of ages and growing sites. Thus, the sequence of remeasurements in time is substituted by simultaneous *point* measurements in space. This method has been used extensively during the 19th century (Kramer, 1988, p. 97; Assmann, 1953; Wenk et al., 1990, p. 116)[1].

Temporary plots are still being used today for constructing growth models in situations where permanent plot data are not available (Lee, 1993; Biber, 1996). The principle is illustrated in Fig. 1-5. Plots of different ages are separated by a vertical line. The x-axis describes in a simplified manner the tree positions, while the symbol t indicates the time axis.

It is possible to reconstruct the development of a state variable that is not affected by competition using point observations and stem analysis. However, the reconstruction of some variables may represent a problem. To explain variations in diameter growth, for example, it is necessary to reconstruct the neighbourhood constellation in the immediate vicinity of the tree. In plot 3, in the left part of Fig. 1-5, for example, a previous competitor carries a question mark, as nothing is known about the tree, except for the remains of a stump.

---

[1] During the 19th century, the *Weiserverfahren* and the *Streifenverfahren* were the most popular methods for obtaining growth information rapidly (Kramer, 1988, p. 97). In the approach known as *Weiserverfahren* the growth of single trees was reconstructed using stem analysis techniques (Hartig, 1868). Another method known as the *Streifenverfahren* was used for gathering data in numerous normally stocked temporary plots of different ages and site qualities to develop yield tables (Baur, 1877).

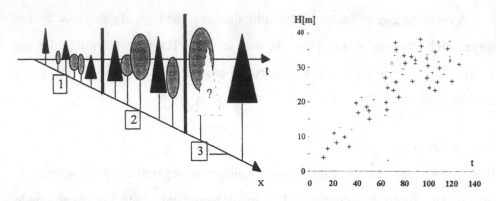

Figure 1-5. Left: Three temporary plots of varying age; the x-axis signifies the tree position; the symbol t indicates the time axis (after Biber, 1996, p. 27). Right: Independent height-age data obtained from temporary plots.

Yield tables were developed in a number of European countries after World War II using data from temporary plots (Pardé, 1961). Examples are the tables used in Great Britain (Hamilton and Christie, 1971), France (Vanniere, 1984) and Spain (Madrigal et al., 1992). These tables are, however, static in nature. They represent the development of a standard silvicultural treatment and cannot be used to predict forest development for alternative thinning regimes (Alder, 1980). The main limitation of temporary plots is the fact that they cannot provide information about the rate of change of a known initial state. Thus, some of the more effective contemporary techniques of growth modelling using a system of differential equations cannot be applied (Garcia, 1988).

Interval plots

A compromise may be achieved by using a system of growth trials which maintains the advantages of permanent plots, i.e. obtaining rates of change of known initial states, as well as temporary plots, i.e. broad coverage of initial states and minimum delay until suitable data are available.

Interval plots are measured twice, the interval between the measurements being sufficiently long to absorb the short-term effects of abnormal climatic extremes. In most cases, an interval of 5 years would be appropriate. The interval is a period of undisturbed growth. Silvicultural operations are not permitted to take place between the two measurements. Measurements should coincide with a thinning operation, to obtain data not only about tree growth, but at the same time about the change of state variables resulting from a thinning. The thinning effects may be assessed at the initial (t_1) or at the final (t_2) measurement, or at both occasions. The concept is illustrated in Fig. 1-6.

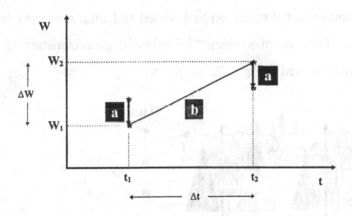

Figure 1-6. Two successive measurements for obtaining the change of state variables resulting from a) a thinning and b) natural growth.

Garcia (1988) proposed a multi-dimensional system of differential equations, in which the future development of a forest depends solely on the present state. To be able to develop such a model, it is necessary to have data describing initial states as well as the associated changes of the state variables. Another suitable model type utilizing the data from two successive measurements is the *algebraic difference form of a growth function*, which has been used by a number of authors (see, for example, applications by Clutter et al. 1983; Ramirez-

Maldonado et al., 1988; Forss et al., 1996). One *algebraic difference form* of Eq. 1-1, for example, may be written as follows:

$$H_2 = H_1 \cdot \left[\frac{1 - e^{-a_1 \cdot t_2}}{1 - e^{-a_1 \cdot t_1}} \right]^{a_2}$$

1-2

with H_1, H_2 = dominant stand height at age t_1 and t_2,
       a_1, a_2 = empirical parameters.

Interval plots, measured twice and spread over a range of growing sites, development stages and silvicultural treatment categories, combine the advantages of the permanent and the temporary designs. Thus, the rates of change of known initial states can be assessed and, after minimum delay, suitable data are available for modelling. The essential characteristics of the interval concept are illustrated in Fig. 1-7.

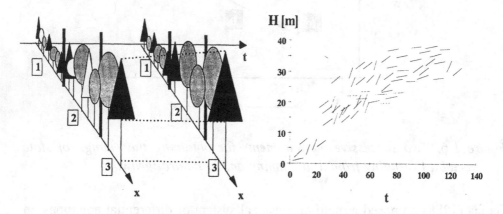

Figure 1-7. Left: Three interval plots; white trees are removed during a thinning operation. Right: Interval data for obtaining rates of change of observed state variables.

The interval concept offers flexibility, as the plots may be abandoned at any time after achieving the objective to obtain one interval measurement. A permanent infrastructure involving staff for repeated remeasurements is thus not required.

Chapter Two

Projecting regional timber resources

Projecting the development of a regional timber resource in response to a given series of periodic harvest levels is an essential prerequisite for wood procurement planning and forest sector modelling. A forest region is characterized by age class distributions, average yield classes and sometimes average degrees of stocking within age classes, for different tree species. The per-ha standing volume of the timber resource may be estimated for a given age, and this information forms the basis for regional resource projections assuming different harvest scenarios.

An even-aged forest develops in cycles from young to old age during the course of a rotation. We may refer to such a forest as a *cyclic forest*. The cycle begins with the establishment of the young trees and ends with the harvest of the mature ones. Intermediate thinnings may improve the value of the final crop. Productivity is measured in terms of the total volume produced, or the mean annual increment at rotation age (MAI$_R$). Sustained yield control is based on the model of the *Normal Forest*, and the present value of a specified silvicultural program is equal to the sum of the annual net incomes, discounted to the age of establishment.

An alternative type of managed woodland, known as *continuous cover forest*, is not uncommon in Central Europe and in tropical regions. Forest development has no beginning or end. The forest remains in a state of undefined age, oscillating about a specified level of growing stock. Harvesting operations may take place at regular intervals and it is not possible to distinguish between a thinning operation and a final harvest. Age-based measures of forest production and valuation, such as mean annual increment or age-based net present value, cannot be used. The only appropriate variable for assessing production, which is difficult to estimate in advance, is the periodic annual increment[1].

In an even-aged forest, however, the available timber volume at a given age may be estimated using an empirical yield function, which is based on observations of growing stock volumes at various development stages. Alternatively, a theoretical or *synthetic* yield function may be used, which has no empirical base other than the estimated mean annual increment (MAI) and the age of culmination of the MAI.

Empirical yield functions

A cyclic forest has a fixed lifetime defined by the rotation. It may develop unthinned, or with intermediate thinnings between planting and final harvest. Unthinned cyclic forests are often found in short-rotation pulpwood plantations or in natural forest areas where access is difficult. Thinned cyclic forests are common in many kinds of commercially-managed woodlands where the quality of the final crop is an important consideration. Obviously, the methods of yield forecasting are much simpler in fully-stocked, unthinned cyclic forests than the methods required for forests with variable degrees of stocking.

---

[1] The adjectives *monocyclic* and *polycyclic*, which are sometimes used to distinguish between a cyclic forest and a continuous cover forest, are misleading. *Polycyclic* development is not only found in continuous cover forests, but also in thinned cyclic forests.

Fully stocked forests

Yield tables that represent the *normal* development of a forest are a frequently used data base for regional timber resource forecasting. A yield table estimates the production potential for a discrete number of site quality classes, assuming average silvicultural treatment. An example of this approach is given by Smaltschinski (1997, p. 132), who used the yield tables developed by Schober (1957), to model the volume of the remaining stand as a function of site class and age (the volume of the *remaining stand* is an expression relating to yield table terminology; it is the portion of the total volume production remaining after thinnings). First, the volume at age 100 is derived from the site class using a parabola:

$$V_{100} = a + b \cdot H_{100} + c \cdot \left(H_{100} \right)^2 \qquad\qquad 2\text{-}1$$

with H_{100} = mean height at age 100 [m],
V_{100} = remaining stand volume at age 100 [m³/ha],
a,b,c = model parameters.

V_{100} is then used to estimate the volume of the remaining stand for a given age and V_{100}, using Eq. 2-2.

$$V_{rem} = \alpha \cdot \left(\frac{V_{100}}{\alpha} \right)^{\displaystyle e^{\gamma} \cdot \frac{\beta}{e^{\gamma}} \left[t^{\gamma} - 100^{\gamma} \right]} \qquad\qquad 2\text{-}2$$

with V_{rem} = volume of the remaining stand [m³/ha],
V_{100} = remaining stand volume at age 100 [m³/ha],
e = base to the natural logarithm,
α, β, γ = model parameters.

The parameters relating to Eq. 2-1 (a, b, c) and Eq. 2-2 (α, β, γ) were calculated for a number of species. They are listed in Tab. 2-1 for beech (*Fagus sylvatica*) and oak (*Quercus robur*).

Species	a	b	c	α	β	γ
Beech	-156,5	25,1	0,062	1604	-13,5	-0,67
Oak	-121,1	19,5	0,056	954	-4,0	-0,40

*Table 2-1. Parameters a, b, c and α, β, γ of equations 2-1 and 2-2, for beech
(fagus sylvatica) and oak (Quercus robur), respectively.*

The method may be used to estimate the development of the volume of the
remaining stand over age for different site indices in beech forests managed
subject to the normal silviculture as defined by the yield table (Fig. 2-1).

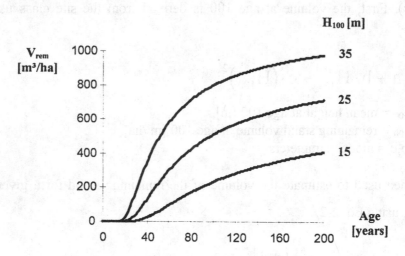

*Figure 2-1. Development of the volume of the remaining stand V_{rem} over age for
three different site indices in beech according to Smaltschinski (1997).*

Another example of an empirical yield function that can be used for regional
resource forecasting, is provided by Shvidenko et al. (1995). Consider the
following yield model for a fully-stocked *Pinus sylvestris* forest growing on a
uniform site of quality class III in northern Russia:

$$V = 205.3 \left[1 - e^{-0.0231 \cdot t} \right]^{2.93} \tag{2-3}$$

with   V = expected volume of growing stock [m³/ha] at the specified age [years].

Equation 2-3 may be used to compile a yield table, such as the one presented in Tab. 2-2.

	Age Class Midpoint									
	10	30	50	70	90	110	130	150	170	190
m³/ha	2.0	26.9	67.7	107.4	138.8	161.4	176.8	187.1	193.7	197.9

Table 2-2. Yield table for a fully-stocked Pinus sylvestris forest growing on a uniform site of quality class III in northern Russia.

Assuming a given initial age-class distribution, a constant periodic harvest of 30,000 m³ per 20-year cutting period and immediate regeneration of the felling areas, the age class distributions may be calculated for a series of future cutting periods (Tab. 2-3).

	Age Class Midpoint										
	10	30	50	70	90	110	130	150	170	190	Total
a_{i0} (ha)	20.0	300.0	100.0	100.0	200.0	50.0	50.0	100.0	10.0	70.0	1000
a_{i1} (ha)	156.0	20.0	300.0	100.0	100.0	200.0	50.0	50.0	24.0	0.0	1000
a_{i2} (ha)	168.3	156.0	20.0	300.0	100.0	100.0	155.7	0.0	0.0	0.0	1000
a_{i3} (ha)	171.0	168.3	156.0	20.0	300.0	100.0	84.7	0.0	0.0	0.0	1000
a_{i4} (ha)	177.7	171.0	168.3	156.0	20.0	300.0	7.0	0.0	0.0	0.0	1000
a_{i5} (ha)	185.2	177.7	171.0	168.3	156.0	20.0	121.8	0.0	0.0	0.0	1000

Table 2-3. Simulated development of the age-class distribution of a Pinus silvestris forest growing on a uniform site of quality class III in northern Russia, and assuming a harvest rate of 30,000 m³ per 20-year cutting period.

The development of the resource presented in Tab. 2-3 shows the effect of a given harvest level on the age-class distribution. In the specific example, a periodic harvest of 30,000 m³ would already result in a wipe-out of the three oldest age-classes in the second cutting period. The normal age-class area would amount to 200 ha, if the rotation of maximum MAI of 100 years is selected. With a mean annual increment of 1.54 m³/ha/year at 90 years, a harvest level of

30,000m³ per 20-year period could be sustained indefinitely (neglecting the risks of fire and other disturbances).

A regional yield model is a useful tool for evaluating the effects of different harvest levels on a given age-class distribution and a simple age-class simulation is often the only feasible way to predict the dynamic development of a forest resource on a regional basis. The method involves, however, considerable aggregation over growing sites, forest types and management regimes.

Figure 2-2. Sparsely-stocked boreal forest (photo A. Dohrenbusch).

Non-fully Stocked Forests

More refined methods of simulation need to be applied in regions where intensive production forestry is practiced, and the first step towards refinement involves a method for considering the effects of different levels of stand density. The density of a forest is often expressed using the variable *degree of stocking*, which implies the observed stocking (basal area or volume per ha) as a proportion of some normal yield table stocking.

The reduction factor *degree of stocking* multiplied by the normal yield table volume gives the initial growing stock volume V_1. A suitable yield model for projecting the volume growth of a forest with a known or assumed initial growing stock per ha is the algebraic difference form of Eq. (2-3), namely:

$$V_2 = V_1 \left[\frac{1 - e^{-\alpha_2 \cdot t_2}}{1 - e^{-\alpha_2 \cdot t_1}} \right]^{\alpha_3} \qquad \text{2-4}$$

where V_1, V_2 = the volumes per ha at ages t_1 and t_2,
  α_2, α_3 = empirical parameters which are a function of site quality and stand density.

Shvidenko et al. (1995) proposed the following two equations for estimating the parameters of Equation 2-4 for a *Pinus sylvestris* forest in Russia:

$$\alpha_2 = \frac{1}{100} \left[0.034 \cdot SQ^2 - 0.430 \cdot SQ + 0.362 Bo^2 - 2.136 Bo + 0.113 \cdot SQ \cdot Bo + 4.285 \right] \qquad \text{2-5}$$

$$\alpha_3 = \frac{1}{10} \left[0.389 \cdot SQ^2 - 2.872 \cdot SQ + 8.230 \cdot Bo^2 - 12.963 \cdot Bo + 0.776 \cdot SQ \cdot Bo + 27.684 \right] \qquad \text{2-6}$$

where SQ = site quality (SQ=1, 2, .., 7),
    Bo = degree of stocking (Bo=0.3, 0.4, ...,1.0).

The degree of stocking is reduced by thinnings. Thus, using Eq. (2-4), with the parameters estimated by (2-5) and (2-6), it is possible to calculate the growth of a stand following a thinning. The algebraic difference form of a volume growth function is an effective tool for projecting a known initial state. However, in the absence of an asymptote, care should be taken with the time horizon, and valid estimates are limited to short-term projections.

Example: The growing stock volume of a 70-year old fully-stocked pure pine stand growing on a quality III site, amounts to 100 m³/ha. Estimate the growth of the residual growing stock during the next 10 years. 20% of the volume will be removed immediately during a thinning operation. Solution: The parameter values after the thinning are α_2=0.021 and α_3=1.9328, and the growing stock volume at age 80 is estimated to be equal to 89.0 m³/ha. The increase of the residual growing stock is thus 89-80=9m³/ha during the next 10 years.

Projections based on yield tables need to be adjusted for variable density. This is generally done using tables of reduction factors (Kramer and Akca, 1987) or specific growth reduction functions for adjusting current increment in accordance with the actual degree of stocking (Smaltschinski, 1997, p. 155). Lemm (1991, p. 92 et sqq.) used the following equation

$$C = Bo^b \cdot e^{b(1-Bo)}$$

2-7

with   C = reduction factor for current increment (basal area or volume)
       Bo = degree of stocking (0≤Bo≤1.0),
       b = parameter (1.97 spruce; 1.81 pine,larch; 1.13 beech,ash,maple; 2.07 oak)

A graph of the relationship between the reduction factor for current increment and the degree of stocking is shown in Fig. 2-3, for beech and spruce forests in Switzerland.

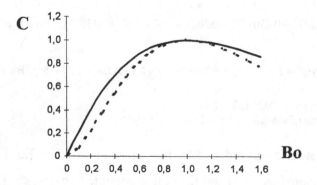

Figure 2-3. Reduction factor for current increment (C) in beech (continuous line) and spruce (broken line) for different degrees of stocking (Bo) according to Lemm (1991).

Yield Functions based on MAI Estimates

Mathematical models for estimating timber yields are usually developed by fitting a suitable equation to observed data. The model is then used to predict yields for conditions resembling those of the original data set. It may be accurate for the specific conditions, but of unproven accuracy or even entirely useless in other circumstances. Thus, empirical yield models tend to be specific rather than general.

Regional yield forecasts are based on aggregate area information, typically in the form of age-class distributions, involving a considerable lumping of area detail with an associated uncertainty. General yield models, covering all possible growing sites and silvicultural regimes, are simply not available. Such models are, however, often required for timber resource projections based on age-class information. In the absence of such models, there are a number of options. One of them is to use aggregate estimates of the mean annual increment (MAI), based on experience or historical data. As tree growth is not linear, such MAI estimates have to be used in conjunction with a reference age. It is convenient to use the age at which the MAI culminates, designated as t_{max}, and the MAI at this age, MAI_{max}. Murray and Gadow (1993) proposed a general yield model, characterized by a variable age of maximum MAI, which is applicable in situations where reasonable estimates of MAI_{max} and t_{max} are available. A suitable yield model is the Chapman-Richards-function:

$$V = A \left[1 - e^{-k\{t-to\}} \right]^m \qquad \text{2-8}$$

The variable t represents time after planting, or age. V is the growing stock of a forest of age t, measured in m^3/ha. The parameter A, the asymptote, is a scaling factor for the size of the variable, its units are the same as those of the dependent variable. The parameter k scales the time axis. By varying k, the model can match the actual growth rate of the variable. The parameter t_0 indicates the lower bound age, after which an increase of the growing stock volume of the forest may be observed. The fourth parameter m offers further flexibility concerning the shape of the growth curve.

Specific values for A, k, t_0 and m are obtained if Equation 2-8 is fitted to empirical yield data. Setting $t_0=1$ is quite satisfactory for fast-growing timber plantations, especially when the emphasis is on making good predictions in the vicinity of t_{max}. The parameter A calibrates the model and its value is automatically determined whenever calibration data are provided, such as by specifying a value for MAI and a corresponding felling age. The form and orientation of the model with respect to the time axis is determined by the parameters k and m. The value of t_{max} is thus exactly defined by the values of k and m. For example, t_{max} is exactly equal to 25 years when k=0.095 and m=3.70. Consequently, k and m cannot be used in a model in which t_{max} is specified as an input variable.

However, a more practical solution may be to specify t_{max}, rather than derive it on the basis of a model fitted to a given data set. The values of MAI_{max} and t_{max}, *MAI* and *MAI culmination age*, are often well-known in practice and it would be practical to define a yield function in such a way that the parameters are in compliance with the known MAI_{max} and t_{max}.

The condition that the MAI culminates at a given t_{max}, requires the derivative of MAI(t), or of $(1-e^{-k(t-1)})^m/t$, to vanish at t_{max}. The multiplicative factor A can be ignored and, upon setting the derivative equal to zero, one obtains the condition:

$$1 = \left[1 + t_{max} \cdot m \cdot k\right] e^{-k\{t_{max}-1\}}$$

2-9

relating the specified value of t_{max} and the parameters k and m. Murray and Gadow (1993) argue that 3 would be an appropriate value for m. Thus, the only remaining parameter k can be determined iteratively.

The yield function based on MAI requires as inputs the known values of t_{max} and MAI_{max}, the former specifying the MAI-culmination age and the latter calibrating the MAI curve. It is therefore convenient to introduce the relative MAI defined by the condition $MAI_{max} = 1$. The relative MAI, now a dimensionless quantity per ha per year, is given by:

$$relMAI(t) = \frac{t_{max}}{t} \frac{\left(1 - e^{-k\{t-to\}}\right)^m}{\left(1 - e^{-k\{t_{max}-to\}}\right)^m}$$ 2-10

Fig. 2-4 shows examples of relative MAI curves culminating at 20, 50 and 80 years.

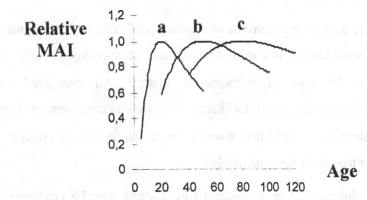

Figure 2-4. Three relative MAI curves (a: $t_{max} = 20$, $k=0.1043$; b: $t_{max} = 50$, $k=0.0394$ and c: $t_{max} = 80$, $k=0.02434$)

The actual MAI curves are obtained by multiplication with the given MAI_{max}. The growing stock for a given age class is equal to the MAI multiplied by the age class midpoint.

Chapter Three

Modelling stand development

Most managed forests are conveniently subdivided into a discrete number of geographical units known as stands. A stand may be even-aged or uneven-aged, depending on the type of management history. An even-aged stand is characterized by one age, which facilitates modelling its development. Even-aged stands are usually, though not always, more uniform with regard to the distribution of tree diameters and heights.

The development of an even-aged forest stand may be predicted using a stand model. Important variables in a whole stand model are the average or dominant stand height, the basal area and the stems per ha. These basic quantities are used to derive secondary values, such as the quadratic mean diameter or the stand volume. Stand volume is sometimes predicted directly.

Height

The average or dominant stand height is an important variable in even-aged forests. Height growth is relatively independent of stand density and thus not much affected by thinning. For this reason, height is not only used to characterize stand development, but also to estimate the potential of the growing site. The growth potential of a given site is measured by the *site index*, the dominant height at a given reference age. Height models and site index equations feature very prominently within the framework of a stand model.

The development of the stand height may be described using a 3-parameter asymptotic function, such as the Chapman-Richards generalization of Bertalanffy's growth model (Pienaar and Turnbull, 1973):

$$H = a_0 \cdot \left[1 - e^{-a_1 \cdot t} \right]^{a_2} \qquad \qquad 3\text{-}1$$

with   H   =   dominant stand height [m]
       t   =   stand age [years]
       a_0 .. a_2 =   model parameters.

The scale parameter a_0, which is expressed in terms of the units of the dependent variable H, defines the asymptotic height. The parameter a_1 scales the time axis, while a_2 offers further flexibility concerning the shape of the growth curve.

The *site index* (SI) is obtained by substituting the variable t in (3-1) with the site index reference age. For example, the site index with reference age 100 is defined by (3-2).

$$SI_{100} = a_0 \cdot \left[1 - e^{-a_1 \cdot 100} \right]^{a_2} \qquad \qquad 3\text{-}2$$

Stand height and age are used to estimate the site index and, for this purpose, it is convenient to use a specific form of (3-1). Both (3-1) and (3-2) may be solved for a_0, and the right-hand sides of both equations are thus equal to a_0. They can be equated and re-arranged to give (3-3), the algebraic difference form of (3-1).

$$SI_{100} = H \cdot \left[\frac{1 - e^{-a_1 \cdot 100}}{1 - e^{-a_1 \cdot t}} \right]^{a_2} \qquad \qquad 3\text{-}3$$

Equation (3-3), when solved for H, may be used to describe the height development over age for a given site index. It may be used specifically to draw a system of site index curves, for example, with a spreadsheet program.

There are a number of univariate growth functions with asymptotes and inflection points, suitable for modelling the development of height over age (Kiviste, 1988; Zeide, 1993; Shvets and Zeide, 1996; Garcia 1997). Several of the available 3-parameter models will give equally good results. However, in the practice of forest modelling, a distinction is usually made between two types of height models, based on the nature of the families of curves they generate. For any two curves, in an *anamorphic* family of height curves, the height of one at any age is a constant proportion of the height of the other at the same age. This proportionality does not hold in a *polymorphic* family of height curves.

Anamorphic height models

A rather simple, but nevertheless useful, height model that has been successfully applied by various authors, is the Schumacher-function (3-4)[1].

$$H = a_0 \cdot e^{-a_1 \frac{1}{t}}$$ 3-4

with   H   =   dominant stand height [m]
         t   =   stand age [years]
         a_0 ,. a_1 =   empirical model parameters.

The two parameters of the Schumacher function may be estimated from height/age data point values. An *"average"* curve for the available data, the *guide curve*, is obtained after substituting the estimated values of a_0 and a_1 in Equation (3-4).

Because of its simplicity, the model is well-suited to demonstrate the difference equation method of fitting a height model based on growth intervals. Fig. 3-1 gives a geometric interpretation of the problem, according to Clutter et al. (1983, p. 51). Point A represents the initial measurement values with coordinates $[1/t_1; \ln (H_1)]$. Point B has coordinates $[1/t_2; \ln (H_2)]$, obtained through remeasurement after a time interval of $(t_2 - t_1)$ years.

---

[1] *vide* Schumacher (1939). In Eastern and Central Europe the Schumacher-function is known as the Michailoff function, according to Michailoff (1943).

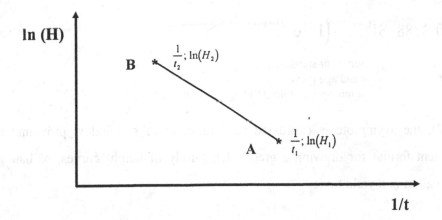

Figure 3-1. Coordinates of the first (A) and second observation (B) on a Schumacher growth curve, according to Clutter et al. (1983, p. 51).

Both points are situated on a line with slope α_1, which is given by:

$$a_1 = \frac{\ln(H_2) - \ln(H_1)}{(1/t_2) - (1/t_1)} \qquad\qquad 3\text{-}5$$

Equation (3-5) is equivalent to (3-6) which can be used to estimate stand height from site index and age, or to calculate site index based on an observed height value and stand age.

$$\ln(H_2) = \ln(H_1) + a_1\left(\frac{1}{t_2} - \frac{1}{t_1}\right) \qquad\qquad 3\text{-}6$$

Based on stem analysis or remeasurement data, the parameter α_1 may be estimated using a linear regression with the model $Y = \alpha_1 X$, where $Y = \ln(H_2) - \ln(H_1)$ and $X = 1/t_2 - 1/t_1$.

A second example of an anamorphic height model is the one developed by Rodriguez Soalleiro (1995, p. 213) for *Pinus pinaster* stands growing in the coastal regions of Galicia in northern Spain. The guide curve was developed using a variant of the Chapman-Richards model proposed by Biging and Wensel (1985):

$$H = 0.2288 \cdot SI^{0.9433} \cdot \left(1 - e^{-0.05493 \cdot t}\right)^{1.4061}$$

3-7

with   H   =   dominant stand height [m]
        t   =   stand age [years]
       SI =   dominant stand height at age 20 years [m].

In (3-7), the asymptote is expressed as a function of site index, providing a convenient format for drawing a graph of a family of height curves, or listing height values in a yield table.

Figure 3-2. A stand of Eucalyptus grandis trees, Westfalia Estate, South Africa.

Disjoint Polymorphic Height Models

The anamorphic approach assumes that the model parameters remain constant for any growing site, i.e. that the shape of the growth curve is independent of environmental factors. This assumption is abandoned in a polymorphic model of height growth, where the form and orientation of the growth curves with respect to the time axis is determined by the site index (Stage, 1963; see Fig. 3-3).

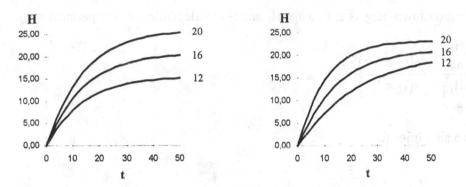

Figure 3-3. Two height models with growth curves for site indices 12, 16 and 20 and a reference age of 20 years. The system on the left is anamorphic, the system on the right polymorphic.

A polymorphic height model can be developed on the basis of long-term growth trials or stem analysis data, providing a set of parameters for each site index. An example of such a model was developed by Jansen et al. (1996) for beech (*Fagus sylvatica*). Expressed as an algebraic difference form of the Chapman-Richards function (Eq. 3-1), their model may be written as follows:

$$H_2 = H_1 \cdot \left[\frac{1 - e^{-a_1 \cdot t_2}}{1 - e^{-a_1 \cdot t_1}} \right]^{a_2} \qquad\qquad 3\text{-}8$$

with H_1, H_2 = dominant heights corresponding to ages t_1 and t_2.
        α_1 = 0.008323+0.0003241 • SI, where SI=dominant height at age 100,
        α_2 = 1.333.

In deriving the difference form, one of the three parameters is constrained using the known initial stand variables, which leaves two parameters to be estimated. In the case of the Chapman-Richards model, it may be preferable to maintain the asymptote and constrain the parameter α_2. This may be achieved by using the logarithmic form of Eq. 3-1 (Amaro et al., 1997):

$$\ln\left(\frac{H}{a_0}\right) = a_2 \cdot \ln\left(1 - e^{-a_1 \cdot t}\right) \qquad\qquad 3\text{-}9$$

The ratio of two heights at two ages t_1 and t_2 is independent of the parameter α_2:

$$\frac{\ln\left(\dfrac{H_2}{a_0}\right)}{\ln\left(\dfrac{H_1}{a_0}\right)} = \frac{\ln\left(1 - e^{-a_1 \cdot t_2}\right)}{\ln\left(1 - e^{-a_1 \cdot t_1}\right)}$$

which simplifies to

$$H_2 = a_0 \left(\frac{H_1}{a_0}\right)^{\frac{\ln\left(1-e^{-a_1 \cdot t_2}\right)}{\ln\left(1-e^{-a_1 \cdot t_1}\right)}} \qquad\qquad 3\text{-}10$$

Equation 3-10, having an asymptote, is more appropriate for use in long-term projections than 3-8. The value of the asymptote parameter α_0 may be determined as a function of site index (Rodriguez Soalleiro, 1995) or used directly as site index (Jansen et al., 1996, p. 8). A polymorphic form is obtained by estimating the shape parameter α_1 from site index, as in the case of Eq. 3-8.

Another example of a polymorphic height model is presented by Rojo and Montero (1996) for *Pinus silvestris* stands in Spàin, using a technique known as *sectioning*[2]. First, the parameters of the Chapman-Richards function are determined for the poorest and for the best sites[3]. In the case of temporary plot data, this approach requires assigning plots to the best and to the poorest sites.

---

[2] The height range is subdivided into a number of equally wide sections; in Germany this technique is known as the Streifenverfahren (Baur, 1877; Sterba, 1991, p. 71).
[3] with $\alpha_0 = (19.962)$, $\alpha_1 = (-0.02642)$ and $\alpha_2 = (1/0.46)$ for $SI_{100} = 17$ and $\alpha_0 = (31.83)$, $\alpha_1 = (-0.03431)$ and $\alpha_2 = (1/0.3536)$ for $SI_{100} = 29$.

The height values for the intermediate sites are obtained by interpolation between the two fitted curves. An example of such a polymorphic system is shown in Fig. 3-4. The height values of site index 29 and 17 representing the fitted curves, are averaged for a given age to obtain the corresponding height values for site index 23. Next, the height values of site indices 29 and 23 are averaged to obtain the corresponding height values for site index 26. This process is repeated for all intermediate height curves.

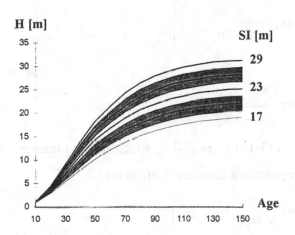

Figure 3-4. Polymorphic height model for Pinus sylvestris *stands in Spain after Rojo and Montero (1996).*

Another height model, more popular in Central Europe, is the Sloboda function which is based on the following differential equation (Sloboda, 1971):

$$\frac{dH}{dt} = \frac{b \cdot H}{t^a} \ln\left(\frac{g}{H}\right)$$
                                                                3-11

with H   =   dominant stand height [m]
      t   =   stand age [years]
      α, β, γ =   empirical model parameters with the conditions α>1, β>0, γ>0.

Equation (3-11) is characterized by some theoretical conditions traditionally applied to a height model (common origin, curve-specific inflection point and asymptote, disjoint growth curves)[4]. The general solution of (3-11) is given by (3-12).

$$H = g \exp\left[-C\exp\left(\frac{b}{(a-1)t^{(a-1)}}\right)\right]$$ 3-12

C is a free parameter and a convenient condition is given by the fact that the height at the chosen reference age (t_o) is equal to the site index (SI). Thus we obtain (3-13).

$$SI = \gamma \exp\left[-C\exp\left(\frac{\beta}{(\alpha-1)t_o^{(\alpha-1)}}\right)\right]$$ 3-13

and

$$-C = \ln\left(\frac{SI}{g}\right)\exp\left(\frac{-b}{(a-1)t_o^{(a-1)}}\right)$$ 3-14

Substituting -C in (3-14) gives (3-12), which may be interpreted as polymorphic as the inflection point is a function of site index:

$$H = g\left(\frac{SI}{g}\right)^{\exp\left(\frac{-b}{(a-1)t_o^{(a-1)}}+\frac{b}{(a-1)t^{(a-1)}}\right)}$$ 3-15

Equ. (3-15) may be used to describe the height development of a stand with a given site index, or to produce a graph of site index curves. For calculating site index directly from age and dominant height Equation (3-15) may be expressed in the following form:

$$SI = g\left(\frac{H}{g}\right)^{\exp\left(\frac{-b}{(a-1)t^{(a-1)}}+\frac{b}{(a-1)t_o^{(a-1)}}\right)}$$ 3-16

The following parameter estimates for even-aged stands of *Cunninghamia lanceolata* growing in the Jiangxi region of China were obtained on the basis of

---

[4] Height curves are not always disjoint as will be shown in the next section.

permanent growth trials complemented with stem analysis data: α = 1.1234; β = 0.3026; γ = 443.3. The resulting system of site index curves is shown in Fig. 3-5.

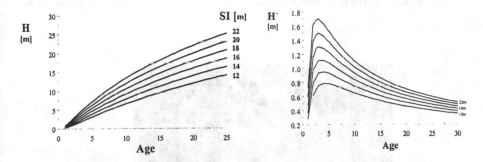

Figure 3-5. Site index system for stands of Cunninghamia lanceolata growing in the Jiangxi region of China (left) with corresponding development of current height growth (right).

Fig. 3-5 shows a culmination of height growth between the ages of 3 and 5 years. This result corresponds with the empirical evidence presented by Yu (1982, p. 37). In accordance with general experience, height growth culminates sooner on the better sites, but declines more rapidly on the good sites after reaching the point of culmination (Kramer, 1988, p. 55).

Another example of a polymorphic height model is provided by Dagnelle et al. (1988) for spruce (*Picea abies*) stands in the Ardenne region of Belgium using the following function:

$$H = a_0 \left(1 - e^{-\frac{t-a_2}{a_1}} \right)^2 \qquad 3\text{-}17$$

with   H = dominant height at age t,
             α_0=2.563+1.73 • SI, where SI=dominant height at age 50,
             α_1=63.39,
             α_2=50 – 63.4 · $\sqrt{-\ln(1 - SI / \alpha_0)}$

The parameter α_1 is a constant, but α_2 is a function of site index. Eq. (3-17) therefore, also belongs to the class of polymorphic height systems.

A prerequisite for developing a polymorphic height model is the availability of suitable height-age data. The data are obtained either from stem analyses or from permanent growth trials which have been observed at least up to the site index reference age. Sufficient site coverage is also desirable, including the best and the poorest growing conditions.

Figure 3-6. A stand of Pinus sylvestris in the Sierra da Guadarrama, Spain.

Non-disjoint Polymorphic Height Models

A rather different approach to developing polymorphic height models was adopted by Kahn (1994), who defined the parameters α_0 and α_1 of Equation (3-1) as a function of 9 site variables, including mean temperature, total precipitation during the vegetation period and nutrient status. The method is not restricted by the availability of quantitative data. Qualitative variables (*good, poor, high, low, etc.*) which are routinely assessed during site surveys, are also included.

The influence of the environmental variables on height growth is assessed using a unimodal trapezoidal transformation function $\mu(X)$, where x describes the value of a given site variable X, and $\mu(X)$ its effect on the height growth. Fig. 3-7 shows a graph of a linear transformation function based on data about the maximum possible height growth of spruce stands growing on sites of different moisture classes in Germany.

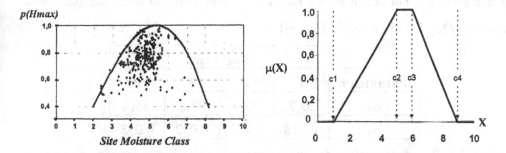

Figure 3-7. Graph of a unimodal trapezoidal transformation function (right) based on data about maximum possible height growth of spruce stands on sites of different moisture classes (left, according to Schübeler et al., 1995). p(Hmax) is the maximum attainable stand height on a given site, relative to the overall maximum height (approx. 42 m).

The transformation function in Fig. 3-7 has 4 parameters (1, 5, 6, 9) defining the five sections of its domain. The parameters are defined separately for each site variable and each tree species, on the basis of data about the specific effects of a given site variable. The function may be described as follows:

$$m(x) = \begin{cases} \dfrac{x - c1}{c2 - c1}, & c1 \le x < c2 \\[2ex] 1.0, & c2 \le x < c3 \\[2ex] \dfrac{c4 - x}{c4 - c3}, & c3 \le x \le c4 \\[2ex] 0, & \text{otherwise.} \end{cases}$$

3-18

with $\mu(x)$ =   effect of x on height growth
           x   =   value of site variable X
c1, c2, c3, c4   =   parameters.

Eventually, the effects of the different site factors are related to the parameters of the Chapman-Richards function, using regression techniques. In this way, site-specific height growth curves may be developed for different tree species.

Kahn applied the method to beech, oak, pine and spruce stands growing on a great variety of site conditions. The parameter α_2 of Eq. 3-1 was held constant at a value of 3.0 for all sites. The remaining two parameters were estimated from the 9 site variables. For beech stands growing on four different sites, the parameter values for Eq. 3-1 are listed in Tab. 3-1.

Parameter	Site Type			
	1	2	3	4
α_0	39.7	36.0	35.2	35.7
α_1	0.018	0.022	0.032	0.033

Table 3-1. Parameter values for the height growth of beech stands growing on four different sites, based on the Chapman-Richards model.

Fig. 3-8 presents a graph of four simulated height curves corresponding to the parameter values listed in Tab. 3-1 for beech forests growing on four different sites.

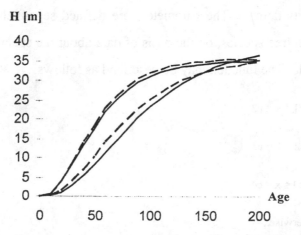

Figure 3-8. Graph of four simulated non-disjoint height curves for beech forests growing on four different sites (according to Kahn, 1994, p. 151).

It is evident from Fig. 3-8 that the height curves are not disjoint, as is usually implied in polymorphic models. The system developed by Kahn (1994) is an example of a polymorphic non-disjoint height model as described by Clutter et al. (1983, p. 57 et sqq.). Such models are not very common. They share the characteristic feature that one or more variables other than height and age are involved.

Basal area

Another important element of a stand growth model is the quadratic mean diameter. This quantity, which is essential for estimating product yields, may be derived from the per-hectare basal area and the number of stems per hectare. Modelling its growth usually proves to be surprisingly easy, given data from permanent plots grown at different densities. However, problems usually arise when the model is applied in thinned stands, where the mean diameter may develop in totally different ways, depending on the type of thinning.

Area-based variables entail the greatest potential for error. For this reason it is preferable, though more difficult, to model basal area and number of stems, and derive the quadratic mean diameter from these quantities using $Dg = \sqrt{40\,000 / \pi \cdot G / N}$. A useful approach is to design the model in such a way that it is possible to project an initially known population of trees. Basal area and the number of stems are of central significance in modelling stand development, the mean diameter, a derived quantity, being of less importance. The final test of a stand model is its ability of predicting area-based variables with high accuracy.

Modelling basal area growth and mortality requires data from plots of varying age and density, remeasured at least once. Temporary plots are inadequate.

Hui and Gadow (1993a) developed the following equation for projecting a known basal area for stands of *Cunninghamia lanceolata* of varying density growing in the southern region of China:

$$G_2 = G_1 N_2^{1-0.142 \cdot H_2^{0.601}} N_1^{0.142 \cdot H_1^{0.601} -1} \left(\frac{H_2}{H_1} \right)^{4.292}$$

3-19

with G_1, G_2 = basal area [m²/ha] corresponding to stand ages t_1 and t_2,
       N_1, N_2 = stems per ha corresponding to stand ages t_1 and t_2,
       H_1, H_2 = dominant stand height corresponding to stand ages t_1 and t_2,

The coefficients in Equation 3-19 were estimated using a large data base of permanent plots from the Fujian and Jiangxi areas. The coefficient of determination (R^2)[5] was 0.979 and the mean square error 0.226. To test for bias, the simultaneous F-test for slope and zero intercept was applied (Hui and Gadow, 1993b; see chapter 6). The F-value of 0.0002, which was obtained for Equation 3-19, is well below the threshold value of $F_{0.05(2,196)} = 3.04$.

Another illustrative example of a basal area model for even-aged stands is provided by Skovsgaard (1997, p. 519) for sitka spruce in Denmark. Rodriguez Soalleiro (1995) developed a basal area increment equation for *Pinus pinaster* stands growing in the coastal regions of Galicia in northern Spain. Per-hectare basal area increment is a function of stand age and stand basal area:

$$\Delta G = 27.78 \cdot G^{0.3367} \cdot t^{-1.3407}$$

3-20

with ΔG = annual basal area growth [m²/ha]
       G = basal area [m²/ha],
       t = stand age [years].

Fig. 3-9 presents a graph of the basal area increment of *Pinus pinaster* stands for different ages and levels of basal area.

---

[5] The coefficient of determination is a ratio indicating how the model compares with a simple average ($R^2=0$) and with the perfect fit ($R^2=1$). It is given by $R^2=1-RSS_m/RSS_\mu$ where RSS_m is the residual sum of squares of the model and RSS_μ is the residual sum of squares about the mean. A high R^2 does not necessarily mean that the model is the best possible one, nor that it will provide good predictions.

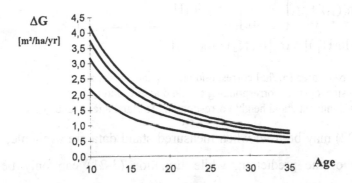

Figure 3-9. Basal area increment for different ages and levels of basal area. The basal area levels in the 4 curves are from top to bottom: 35, 25, 15 and 5 m²/ha.

In stands of the same age and with the same number of surviving trees per hectare, basal area growth will increase with increasing dominant stand height. The model developed by Pienaar et al. (1990) for *Pinus elliottii* plantations in the Atlantic Coast flatwoods of Georgia takes account of this fact:

$$\ln(G) = \alpha_0 + \alpha_1 \frac{1}{t} + \alpha_2 \ln(H) + \alpha_3 (N) \qquad \text{3-21}$$

with   H = dominant stand height [m]
t = stand age [years]
G = basal area [m²/ha] and N = stems per ha,
$\alpha_0, .. , \alpha_3$ = empirical parameters.

Forss et al. (1996) fitted Equation 3-21 to basal area data derived from unthinned stands of *Acacia mangium* grown in South Kalimantan, Indonésia, and obtained the following parameter values: $\alpha_0 = -4.147$, $\alpha_1 = -2.074$, $\alpha_2 = 0.9958$, $\alpha_3 = 0.6239$.

Although Equation (3-21) does not include all of the independent variables in the model of Pienaar et al. (1990), it represents the previously mentioned tendencies in basal area development.

Equation (3-21) may be converted to a basal area algebraic difference form with the same coefficients:

$$\ln(G_2) = \ln(G_1) + \alpha_1 \left[\frac{1}{t_2} - \frac{1}{t_1}\right] + \alpha_2 [\ln(H_2)$$
$$- \ln(H_1)] + \alpha_3 [\ln(N_2) - \ln(N_1)]$$

3-22

with G_1, G_2 = basal area [m²/ha] corresponding to stand ages t_1 and t_2,
N_1, N_2 = stems per ha corresponding to stand ages t_1 and t_2,
H_1, H_2 = dominant stand height corresponding to stand ages t_1 and t_2,

Equation (3-22) may be used when measured stand data are available, adding to the accuracy of the prediction, while Equation (3-21) can only be used to generate idealized basal area curves.

A suitable basal area function may be considered the most important single element of a stand growth model, because it can be used to provide a constraint on individual tree models, and thus form a link between models of high and low resolution.

Potential Density

Estimating the potential density of forest stands, in terms of the surviving trees per hectare, is a central element of growth modelling. It is also one of the most difficult problems to solve, mainly because suitable data from untreated, densely-stocked stands are very rarely found. Furthermore, the natural decline of the number of surviving trees in an unthinned forest is usually characterized by intermittent brief spells of high mortality, followed by long periods of low mortality. The process is not a continuous one, and it is very difficult to model the buildup of susceptability, the phase during which the trees will not succumb, although they may suffer severely, followed by a relatively brief *crash phase*, which leaves large numbers dead (Boardman, 1984; Gadow, 1987, p. 21). Stochastic models have been used in some cases to mimic these processes. However, for the purpose of simulating alternative silvicultural regimes, it is generally assumed that natural mortality is a continuous process.

The Limiting Line

Populations of trees growing at high densities are subject to density-dependent mortality or *self-thinning*. For a given average tree size there is a limit to the number of trees per hectare that may co-exist in an even-aged stand. The relationship between the average tree size (increasing over time) and the number of live trees per unit area (declining over time) may be described by means of a limiting line. A convenient model for this limiting relationship was used by Reineke (1933):

$$N_{max} = \alpha_0 D g^{\alpha_1}$$ 3-23

with   Nmax = maximum number of surviving trees per ha,
    Dg   = quadratic mean diameter [cm],
    α_0, α_1 = empirical parameters.

The parameters of Equation 3-23 can be obtained from fully stocked, unthinned trials, such as the spruce growth series established in Denmark (Skovsgaard, 1997, p. 97 et sqq.) or the *Correlated Curve Trend* (CCT) series of growth experiments established by O'Connor (1935) in South Africa. An example of the limiting line fitted to the data from the unthinned *Pinus radiata* CCT experiment RAD/31 established in Tokai, South Africa, is shown in Fig. 3-10.

The CCT trial is a classic spacing trial designed to predict yields from plantations of various species of pine and eucalyptus for a wide range of densities, varying between extremely dense (2965 stems per ha) and free growth (124 stems per ha). The typical CCT experiment consists of 18 plots, covering 0.04 ha each, some having up to four replications. Nine of the 18 plots were left unthinned, the other nine were subjected to various thinning regimes (Gadow and Bredenkamp, 1992, p. 55 et sqq.). Data from the CCT studies have been used by numerous scientists evaluating the effect of stand density on tree growth and mortality.

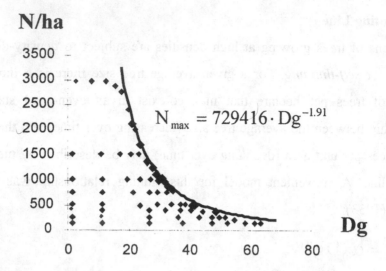

Figure 3-10. Relationship between the average tree diameter (Dg, increasing over time) and the number of live trees per unit area (N, declining over time) for Pinus radiata, CCT experiment RAD/31, Tokai, South Africa.

Considering the high cost of maintaining a series of unthinned, densely-stocked stands, such data are usually not available. To overcome this deficiency, various indirect methods have to be used to estimate the limiting line. Gadow and Hui (1993) compared two different methods for estimating potential density, based on data of unthinned stands of *Cunninghamia lanceolata* from the southern region of China. The first approach follows a procedure developed by Sterba (1975) for estimating the limiting line, the second method uses a differential equation proposed by Clutter and Jones (1980).

Estimating potential density

The relationship between the quadratic mean diameter (Dg), the dominant stand height (Ho) and the number of stems per unit area (N) can be described using a hyperbola (Goulding, 1972):

$$Dg = \frac{1}{\alpha_0 H^{\alpha_1} N + \beta_0 H^{\beta_1}}$$

3-24

The parameters of Equation 3-24 were obtained using data from unthinned *Cunninghamia lanceolata* trials grown in the central region of China. The estimated values are: $\alpha_0 = 0.0000007309$; $\alpha_1 = 0.8932$; $\beta_0 = 1.3806$; $\beta_1 = -1.2324$. The per-hectare basal area is equal to

$$G = \frac{\pi}{4} Dg^2 \cdot N = \frac{\pi \cdot N}{4\left[\alpha_0 H^{\alpha_1} N + \beta_0 H^{\beta_1}\right]^2} \qquad 3\text{-}25$$

Upon setting the first derivative of (3-25) with respect to N equal to zero, the stems per ha at maximum basal area are obtained (Sterba, 1975):

$$N_{Gmax} = \frac{\beta_0}{\alpha_0} H^{(\beta_1 - \alpha_1)} \qquad 3\text{-}26$$

Substituting N in Equation 3-25 with the right hand side of Equation 3-26 gives the quadratic mean diameter at maximum basal area:

$$Dg_{G\,max} = \frac{1}{2\beta_0 H^{\beta_1}} \qquad 3\text{-}27$$

Sterba (1987) eventually achieves the connection with Reineke's model by solving (3-27) for H (assuming that the regression may be inverted) and substituting this expression in (3-25), thus obtaining:

$$N_{G\,max} = \frac{\beta_0}{\alpha_0} \cdot (2\beta_0)^{\frac{\alpha_1}{\beta_1} - 1} Dg_{G\,max}^{\frac{\alpha_1}{\beta_1} - 1} \qquad 3\text{-}28$$

The resulting equation represents the limiting line, with the parameter values estimated for the *Cunninghamia* data, using Equation 3-24.

Based on our experience, this technique for determining the limiting line appears to be suitable for estimating potential density with data from unthinned stands. However, the data set should include stands where self-thinning has occurred, otherwise the slope of the limiting line, which is known to be a rather variable quantity, will be estimated incorrectly. The most famous example of an incorrect slope is the constant −1.605, which was derived simply on the basis of a sample of stands with different mean diameters and stems per hectare (Gadow, 1987).

A suitable test for evaluating the method would involve a subset of data from fully-stocked CCT experiments (or any similar trial). The subset, used to estimate the parameters of the limiting line, should contain data that are not yet affected by self-thinning. It would thus be possible to compare the estimated limiting line with the observed one.

Natural decline of stem number

In the practice of forest management, the limiting line is hardly ever reached (see, for example, Gadow and Bredenkamp, 1992, p. 75). However, this does not mean that trees do not die before the limit is reached. The results from spacing trials show that mortality processes are effective well before the maximum basal area is attained. For this reason, some authors prefer a different approach for modelling the natural decline of stem number.

Clutter and Jones (1980) used a differential equation for modelling the rate of change in the number of stems per ha (N) as a function of stand age:

$$\frac{1}{N}\frac{\partial N}{\partial t} = \alpha \cdot N^{\beta} \cdot t^{\gamma} \qquad\qquad 3\text{-}29$$

with   N = number of surviving trees per ha,
      t = stand age [years] and α, β, γ = empirical parameters.

It is convenient, after integrating 3-29, to use the following algebraic difference form:

$$N_2 = \left[N_1^{\,a} + b\left(t_1^{\,c} - t_2^{\,c} \right) \right]^{\frac{1}{a}} \qquad\qquad 3\text{-}30$$

with N_1, N_2 = stems per ha corresponding to stand ages t_1 and t_2,
      a, b, c = empirical parameters.

Using the data from unthinned *Cunninghamia lanceolata* growth trials, Gadow and Hui (1993) obtained the following parameter values: a=-0.5649; b=-0.0000004423; c=2.3914. The stem number decline was estimated bias-free, and the average relative deviation was 0.07%.

A slightly extended version of (3-30), including site index as an independent variable, was used by Pienaar et al. (1990) for *Pinus elliottii* plantations:

$$N_2 = \left[N_1^a + \left(b + \frac{c}{SI}\right)\left(\left[\frac{t_2}{10}\right]^d - \left[\frac{t_1}{10}\right]^d\right)\right]^{\frac{1}{a}}$$ 3-31

with N_1, N_2 = stems per ha corresponding to stand ages t_1 and t_2.
         SI = site index;
         a, b, c, d = empirical parameters.

Forss et al. (1996) applied the model to the data of *Acacia mangium* stands in Indonesia and obtained the following parameter values: a=-0.1707; b=0.05222; c=-0.5914; d=0.9888. Equation 3-31 can thus be used to generate survival curves for different site indices.

State space models

A state space model defines the rate of change of a forest stand at a given point in time, using several variables that adequately describe its state. García (1984, 1988, 1994) used the three variables dominant height (Ho), basal area (G) and average spacing (i.e. number of trees per hectare, N) for describing the current state space of an even-aged *Pinus radiata* stand. The assumption is that the current location of a stand in this space completely determines its future development, independently of the history of past silvicultural treatments. This concept is very attractive because it does away with the need to design different age-based models for height, basal area and stems per hectare. Indeed, stand age is of no importance in this model, which may be used to evaluate Markov chains (Vanclay, 1994; p. 29 et sq.).

The slopes of the trajectories at any point in the state space are modelled using differential equations. García (1984) used a multivariate version of the Bertalanffy-Richards function. The multivariate version for an n-dimensional state vector x is:

$$\frac{d\mathbf{x}^C}{dt} = \mathbf{A}\mathbf{x}^C + \mathbf{b}$$

3-32

with $\mathbf{x}^C$ defined as $\mathbf{x}^C = \exp[\mathbf{C}\ln(\mathbf{x})]$, where $\mathbf{x}$ is an n-dimensional state vector and $\mathbf{A}$, $\mathbf{b}$ and $\mathbf{C}$ are n-dimensional matrices and vectors of parameters.

The models developed by García (1988) included three to five state variables, with site index entering into the equations in various ways. Site index is a factor multiplying the right hand side of 3-32. Its effect is thus a change in the time scale. In some cases, a reduction in growth was detected immediately following heavy thinnings. To account for this effect, an additional state variable was included, representing a measure of relative site occupancy.

García (1994) presented a two-dimensional state space model derived from a published basal-area increment (Clutter, 1963) and height model (Clutter and Lenhart (1968). The equations were converted to metric units and reformatted as transition functions of current state:

$$\frac{dH}{dt} = 0.0752 \cdot H \cdot \left[3.59 - \ln(H)\right]^2$$

$$\frac{dG}{dt} = 0.0752G \cdot \left[4.08 - \ln(G)\right] \cdot \left[3.59 - \ln(H)\right]$$

3-33

with H = dominant stand height [m] and G = stand basal area [m²/ha].

Fig. 3-11 shows the development trajectory of a hypothetical stand, with an initial height of 5m and an initial basal area of 5m²/ha at age 7. 5m² basal area per ha is removed in a thinning, when the dominant height is equal to 9m. Clearfelling has been scheduled to take place when the stand has reached a dominant height of 25m, which will occur at age 35.

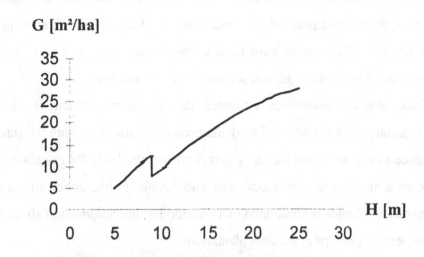

Figure 3-11. Trajectory of a stand with one thinning, based on Eq. 3-33 according to Garcia (1994).

In the state space approach, the fundamental data units for estimating the model parameters are pairs of consecutive measurements, or *interval data*. Suitable interval data are obtained from permanent or semi-permanent growth trials. To obtain data about thinning effects and to ensure undisturbed growth between successive enumerations, measurements of the trials should be planned to coincide with thinnings.

When selecting trial plots, the aim should be to obtain a broad coverage of initial states, including extremes of density, site quality and age. Model performance is determined by the variety of initial states used for estimating the model parameters. For this reason, it might be preferable to establish a new plot instead of remeasuring one for which remeasurement data are already available.

Stand Volume and Product Yields

Estimates of total volume and product yields are an important part of a stand model. Such estimates are indispensable when silvicultural decisions are based on economic criteria. It is impossible to measure the stand volume directly in the field. Therefore, this quantity must be calculated from other variables, such as basal area and dominant height, and sometimes also stand age.

Thus, although sometimes estimated directly, stand volume is often a derived quantity, a *by-product* of modelling, and it is useful to note the relative importance of the measured variables that represent the basis for calculating the volume of a stand, primarily *basal area* and *height*. Furthermore, information about product volumes is considered more important than information about total volumes, except perhaps in biomass plantations.

Figure 3-12. Harvesting 10-year old Pinus patula in South Africa.

A typical example of a stand-level volume equation is given by Rodriguez Soalleiro (1995, p. 213):

$$V = 0.4215 \cdot G \cdot H \qquad \text{3-34}$$

with V=stand volume [m³/ha], G=stand basal area [m²/ha] and H=dominant height [m].

Some authors prefer to use age as a third independent variable. An example of such an approach is the volume equation for Central European forests developed by Jansen et al. (1996, p. 11):

$$V = G \cdot H^{(1.073-0.00133 \cdot t)} \cdot e^{(-1.144+0.00432 \cdot t)} \qquad \text{3-35}$$

with V=stand volume [m³/ha], G=stand basal area [m²/ha], H=dominant height [m] and t=stand age [years].

Product yields may be estimated using a *volume-ratio* approach, such as the one proposed by Amateis et al. (1986). The yield of logs with a minimum thin-end diameter *m* may be estimated using the following model:

$$V_{m,d} = V \left[e^{\gamma_1 \left(\frac{m}{Dg} \right)^{\gamma_2} + \gamma_3 \left(\frac{d_m}{Dg} \right)^{\gamma_4}} \right] \qquad \text{3-36}$$

with $V_{m,d}$= volume of logs with a minimum thin-end diameter *m* [m³/ha],
  V = total stand volume [m³/ha],
  Dg = quadratic mean diameter [cm],
  d_m = minimum Dg yielding products with a thin-end diameter m [cm],
  $\gamma_1 \ldots \gamma_4$ = parameters to be estimated.

The variable d_m is usually estimated from m. Thus, it is possible to draw a curve for a given product type, indicating its relative share of the total volume. The model presented in Fig. 3-13 is based on data from spruce stands grown in Northern Germany, with the following parameter values: $\gamma_1 = -1.311$; $\gamma_2 = 2.877$; $\gamma_3 = -0.1019$; $\gamma_4 = 0.8377$ (the variable m in this specific case refers to a minimum *mid*-diameter of logs, measured in cm, while d_m is estimated by:

$d_m=0.832+0.6688m$). The relative yield proportion of logs with a mid-diameter of between 20cm and 30cm in a stand with a quadratic mean diameter of 34cm, is given by the difference between the 20cm- and 30cm-curves, indicated by the double-sided arrow.

Example: A forester wants to estimate the yield of logs with $20 \le m \le 30$ in a *spruce* stand that is about to be thinned. The total underbark volume removed during the thinning operation amounts to 60m³/ha, and the quadratic mean diameter of the removed trees is equal to 34cm. Using Equation 3-36 we obtain $V_{20,14}=60(0.72) = 43.2$m³/ha and $V_{30,14} = 60(0.38) = 22.8$m³/ha. The estimated log yield amounts to (43.2 minus 22.8) 20.4m³/ha.

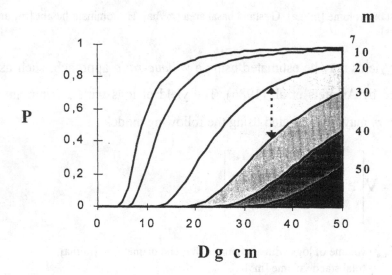

Figure 3-13. Relative proportion P of logs with a minimum mid-diameter m *in spruce stands with quadratic mean diameter Dg, grown in Northern Germany.*

It is sometimes convenient to use a computer program for improved clarity. The following program, *ProdVol*, demonstrates the estimation of the relative share of logs with a mid-diameter of between 20 and 30 cm for a spruce stand with a quadratic mean diameter of 34 cm (see Program 3-1).

```
Program ProdVol;
Var p1, p2: real;

Function power(x, a: real): real;
Begin power:=exp(a*ln(x)) End; {power}

Function P(Dg,{quadratic mean diameter}
            m, {Minimum mid-diameter of log}
            dm {minimum Dg yielding logs}
            :real): real;
Const g1=-1.311; g2=2.877; g3=-0.1019; g4=0.8377;
Begin
    P:=exp(g1*power(m/Dg,g2)+g3*power(dm/Dg,g4));
End; {P}

BEGIN
    {example}
    p1:=P(34,20,14);
    p2:=P(34,30,20);
    writeln(100*(p1-p2):5:1,' percent');
    readln
END.
```

Program 3-1. Program ProdVol *demonstrating the estimation of the relative share of logs with a mid-diameter of between 20 and 30 cm, for a spruce stand with a quadratic mean diameter of 34 cm.*

Thinning Models

The development of a managed forest is not only determined by natural tree growth, but also by the weight and type of the prescribed thinning operations. A thinning reduces the density of a stand and modifies its structure, and thus greatly influences its subsequent medium and long-term development. It may be said that the evolution of a managed woodland is mainly the result of a specific sequence of silvicultural *interferences*.

Whereas growth modelling has always attracted much interest among scientists, and continues to be the subject of numerous research projects, *interference modelling* is not very well developed, which is rather surprising in view of (i) the great impact of silvicultural operations on future stand development and (ii) the relevance of the type of a specific thinning operation for estimating product yields. The commonly used semantic variables for describing different thinning types are not sufficiently precise for accurate predictions of stand development.

The modifications of stand structure, though difficult to model in uneven-aged mixed forests, are more easily defined in even-aged regular stands.

Classical Description of Thinning Operations

In the practice of forest management, thinnings are usually described by indicating the weight and the type of interference. According to the thinning grades prescribed by the German Society of Forest Research Organizations (Wimmenauer, 1902), a moderate low thinning, for example, removes the dead or dying and the suppressed trees, as well as some of the malformed dominant ones (Kramer, 1988, S.180). The adjective *moderate* refers to the weight of a thinning.

Such a prescription can be successfully implemented only if it is possible to identify the trees that are to be removed and distinguish them from those that must remain in the forest. This may be achieved using a tree classification system. Examples of such systems are those developed by Kraft (1884), by the Society of Forest Research Organizations (Wimmenauer, 1902) and the system proposed by the International Union of Forest Research Organizations (IUFRO, 1956).

The purpose of using a tree classification system was to eliminate, as far as possible, the subjective element of foresters engaged in marking trees for removal. Thinning grades based on tree classes are linked to yield tables, which are used in silvicultural planning and forecasting. Some of the terms which are used to describe a standard thinning operation are shown in Fig. 3-14.

Grade		Suppressed Trees[6]			Dominant Trees[7]	
		5	4	3	2	1
A	remain		O	O	O	O
	removed	O				
B	remain			O	◠	O
	removed	O	O		◡	
C	remain					◠
	removed	O	O	O	O	◡
D	remain		O	O	◠	◠
	removed	O			◡	◡
E	remain		O	O	◠	◠
	removed	O			◡	◡

Figure 3-14. Schematic representation to illustrate the thinning grades used in conjunction with the German Yield Tables (Schober, 1994). Semicircle: partially removed/remaining; Full circle: completely removed/remaining.

The aim of a heavy high thinning (*starke Hochdurchforstung*), originally prescribed for old stands, was to favor a limited number of potentially excellent trees (*target trees*), those capable of producing the highest rate of value increment, by removing their immediate competitors (Wiedemann, 1935). First removed were those trees limiting the crown development of a target tree. The method of *high thinning*, applied for example by Heck (1904), Schwappach (1905) and Michaelis (1907) was developed and applied in Germany following the example of the *eclaircie par le haut* practiced in France before the turn of the century (Schober, 1991; Brandl, 1992). Because specific emphasis is placed on the selection of target trees, *high thinning* is also known as *selective thinning*. The classic selection thinning described by Schädelin (1942) and Leibundgut (1978, p. 116 ff.) was originally designed for beech forests, and one of the characteristics of the method is the fact that the best trees (*target* trees) were re-evaluated at each thinning, since the relevant quality attributes of a target tree

---

[6] Class 3: weakly co-dominant; class 4: suppressed; class 5: completely suppressed.

[7] Class 1: dominant with exceptionally large crowns; class 2: co-dominant with well developed crowns.

could change between successive thinning operations. Since the early 1970's, the term *selective thinning* is also used to refer to thinnings in pine and spruce forests. However, in coniferous stands, target trees are selected early and marked permanently (Abetz, 1975; Johann, 1982). Thus *high thinning* and *selective thinning* are two different terms which carry the same meaning, leading to unneccessary misunderstanding and confusion.

Another concept which may give rise to misunderstandings, is the so-called *Plenterdurchforstung*. Borggreve (1891) used the term to refer to the removal of the largest trees (thinning from above), thus enhancing the growth potential of the previously suppressed ones, which were often of superior quality (Brandl, 1992). The same term, however, is used by Swiss foresters to describe a thinning operation that is used to transform a forest into a *Plenterwald*, mainly by removing the so-called *intermediary* or medium-sized individuals (Schütz, 1989).

Numerous efforts have been made over the years to develop a terminology for describing different types of thinnings. Some authors use graphs such as Fig. 3-15 to define a thinning operation.

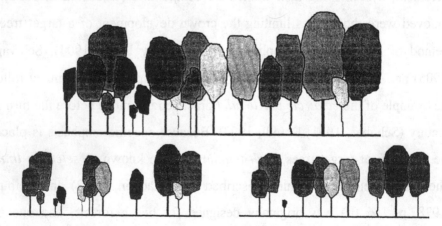

Figure 3-15. Schematic representation of two different thinning operations in an even-aged beech stand. Above: stand before the thinning; below left: after a heavy high thinning; below right: after a moderate low thinning (cf. similar graphs by Kramer, 1988; Dengler, 1982, p.167).

Notwithstanding these efforts, it is hardly possible to give a sufficiently precise description of a thinning that has already been completed, let alone an accurate prescription of one to be carried out in the future. The number of terms that can be used for describing different thinning operations is limited. One of the first authors to recognize this problem was Franz (1972), who introduced a thinning factor for describing simultaneously the type and weight of a thinning.

The increasing differentiation of thinning types has been a distinct feature of European silviculture. The ability to describe a complex phenomenon using simple terminology is limited (cf. Ammon, 1951, p. 50 f). Therefore, an important objective in growth modelling is the development of better tools for describing silvicultural interferences as precisely as possible. Such descriptive tools are a prerequisite for developing prescriptions and forecasts of future interferences. Within the framework of a stand model, it is necessary to use variables defining the weight and others characterizing the type of a thinning.

Thinning Weight

A measure for quantifying thinning weight at the stand level, which is sometimes used in conjunction with a yield table, is the reduction in the degree of stocking $(B°)$[8] caused by the thinning. Kramer (1990) presents a practical method for estimating thinning yields based on the reduction of $B°$.

Another popular measure for defining the thinning weight is the change in *relative spacing* first introduced by Hart (1928), and which is also known as the Hart-Becking spacing index or S-percent. Relative spacing is defined by the average distance between the trees, expressed as a proportion of dominant stand height which, if square spacing is assumed, can be written as follows:

---

[8] The degree of stocking is the basal area per hectare expressed as a proportion of some normal basal area, defined by a yield table.

$$RS = \frac{\sqrt{10000/N}}{H} \qquad\qquad\qquad 3\text{-}37$$

with   RS= relative spacing index,
       H = dominant stand height [m] and N = stems per ha.

Equation 3-37, when used for developing thinning guides where RS is being specified, must be solved for N. Thus $N = 10000/(H \cdot RS)^2$.

Example: The first thinning in a spruce stand planted at 2500 trees per ha will be imposed when the dominant height has reached 10m. The thinning prescription specifies a relative spacing of 0.23 after thinning. The residual number of stems remaining after the thinning is thus equal to $10000/(2.3)^2=1890$ stems per ha, i.e. approximately every fourth tree is removed.

Relative spacing is sometimes used for constructing thinning ranges which define the maximum and the minimum number of trees for a given stand height. A hypothetical thinning range is shown in Fig. 3-16.

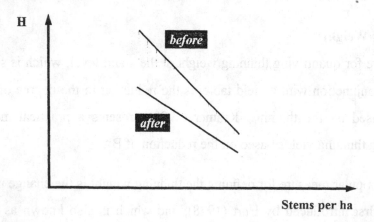

Figure 3-16. Hypothetical thinning range defining the maximum and minimum number of trees for a given stand height.

When a stand has reached the line marked before, it should be thinned to avoid undesirable crowding and associated instability. The difference between the before and after ordinate values for a given height defines the thinning weight, i.e. the stems per ha that should be removed during the thinning. Examples of thinning ranges for even-aged forests are given by Abetz (1975) for European spruce, by Lewis et al. (1976) for Australian *Pinus radiata* and by Gadow and Bredenkamp (1992, p. 61) for pine plantations grown in the summer rainfall areas of South Africa. Specific thinning ranges may be developed to facilitate the planning of silvicultural operations. However, these decision aids should be used with caution because the weight of a thinning does not only depend on the tree species and the silvicultural objectives. Adjustments have also to be made to accomodate abnormal stand conditions such as overcrowding.

Type of Thinning

The future development of a forest is not only influenced by the weight, but also by the type of thinning, which is defined by the selective removal of specific members of the population. Fig. 3-17 shows the effects of a low thinning and a high thinning, in a 0.1 ha sample plot situated in a 60-year old beech stand.

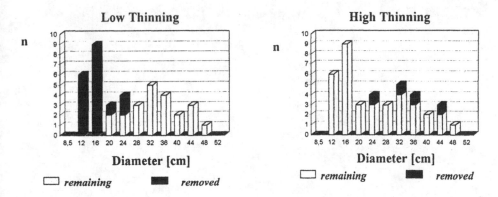

Figure 3-17. Change in the diameter distribution of a 60-year old beech stand, resulting from a low thinning (left) and a high thinning (right).

The weight of the two thinnings was approximately the same, with 15 % of the basal area removed in the low thinning and 16 % in the high thinning. However, the low thinning removed 45 % of the trees, the high thinning only 10 %, the difference being due to the type of thinning.

The type of thinning is also reflected by the change of the diameter distribution. This change may be described in different ways as explained in the next chapter. At the stand level, diameter distributions are usually not available. In this case, the SG-ratio may be used:

$$SG = \frac{(N_{thn} / N_{tot})}{(G_{thn} / G_{tot})}$$

3-38

with N_{thn}, N_{tot} = removed and total number of stems;
           G_{thn}, G_{tot} = removed and total basal area.

Using the data presented in Fig. 3-17, we obtain 0.45/0.15=3.0 for the low thinning, and 0.10/0.16=0.625 for the high thinning.

Chapter Four

Size-class Models

Growth and yield models are needed for practical use within the framework of a forestry planning system. This is easier said than done, considering that it is necessary to project forest development for a broad spectrum of site and treatment conditions, and for different levels of detail. Variations in the level of detail of the available information requires models of varying levels of resolution.

Size-class models project the development of a limited number of trees representing the entire population. Each *representative tree* exhibits the attributes of a subpopulation of trees, which are similar in some way. The groups are formed by subdividing the population into a fixed number of units known as *cohorts*. The trees in a *size-class cohort* belong to a diameter class of equal width. The diameter classes usually contain unequal numbers of trees. The trees in a *percentile cohort* may vary somewhat in size, but each cohort contains the same number of trees. Using a hypothetical cumulative diameter distribution, the difference is illustrated in Fig. 4-1. In uneven-aged mixed forests, cohorts may be formed on the basis of various other attributes, such as species, stem quality or crown size (Leary, 1979; Vanclay, 1994, p. 51 et sqq.).

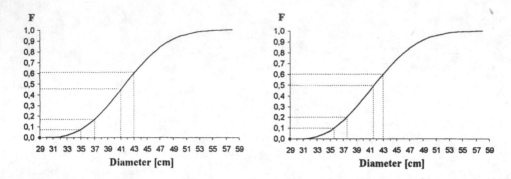

Figure 4-1. Cohorts are formed by subdividing the population into a fixed number of units each comprising trees that are similar in some way. Left: size-class cohorts; right: percentile cohorts.

Thus, a size-class cohort implies unequal numbers of trees in the different size-classes, while a percentile cohort implies size-classes of unequal width. Size-class cohorts are more common, while percentile cohorts, as used for example by Clutter and Allison (1974) and Alder (1979), are not very frequently encountered.

Size-class models represent a compromise between stand models and individual tree models. They expand the computational effort applied in a stand model (which projects the development of one representative tree with average population attributes) and reduce the level of detail required in a single tree model (which maintains the specific attributes of each tree in the population).

Size-class models are essential for estimating product yields. They generate future diameter distributions and *stand tables* based on a known initial diameter distribution. The basic modelling units are the *representative trees*, which embody the specific sizes, shapes and growth rates of all the members of the cohort.

Diameter Growth

Breast height diameter is a convenient variable for predicting the growth of representative trees in a size-class model. Being closely correlated with stem volume and value, diameter is an essential quantity for economic and silvicultural decision-making. Furthermore, the diameter is easy to measure and initial diameter distributions are often available from forest inventories, providing a reliable base for projections.

A number of methods have been developed for projecting diameters of representative trees in a size-class model. Although a classification of these methods is difficult, they can be conveniently listed under the headings *diameter distribution projection* and *stand table projection*.

Figure 4-2. Young pine stand after thinning and pruning, Northern Germany.

Projecting Diameter Distributions

The use of diameter distributions for predicting forest development has a special appeal, and a rather long tradition in even-aged growth modelling (Meyer, 1930;

Prodan, 1953; Kennel, 1972; Bailey and Dell, 1973; Hafley and Schreuder, 1977).

Useful information for forest planning and yield estimation is obtained by breaking down stand variables into diameter classes. Various techniques have been used to estimate the parameters of the diameter distributions, and attempts have been made to classify them. Usually, a distinction is made between the *parameter prediction method* and the *parameter recovery method*. The *parameter prediction* technique estimates the parameters of a future distribution from stand attributes such as dominant height, mean diameter, age and density. In the *parameter recovery* approach, the rates of change of the distribution moments or percentile values at time t are derived from the known moments or percentile values and other stand variables at time t, such as dominant height. In effect, both methods estimate the distribution parameters from known or estimated stand attributes, using empirical functions.

The parameter prediction technique is the more common approach. The first step involves fitting a theoretical diameter distribution to observed diameter frequencies in a variety of stands with given attributes. Statistical relationships are then established between the distribution parameters, and those stand attributes that are thought to be usually available.

One of the most popular distribution functions is the three-parameter *Weibull* model. The cumulative distribution function of X gives the probability that a value less than or equal to X will occur:

$$F(X) = P(x \leq X) = 1 - e^{-\left(\frac{X-a}{b}\right)^c} \qquad\qquad 4\text{-}1$$

with
$\quad$ x $\;=$ random diameter at breast height [DBH, cm],
$\quad$ X $\;=$ DBH, for which the probability is calcuated that x assumes a smaller value,
$\quad$ F(X) $=$ P(x ≤ X) = cumulative probability of the random variable x
$\qquad\quad\;=$ probability that a random DBH is smaller than X
$\quad$ a, b, c $=$ location, scale and shape parameter of the Weibull distribution.

Smalley u. Bailey (1974) presented one of the first applications of the parameter prediction method. Their equations are used to estimate the Weibull parameters for pine plantations in the Tennessee, Alabama and Georgia highlands:

$$a = \begin{cases} -1.9492 + 0.0757 \cdot H, & H \geq 26 \\ \\ 0, & H < 26 \end{cases} \qquad \text{4-2}$$

$$b = -a - 5.2352 - 0.0003 \cdot N + 1.1955(10)^3 / N + 6.2046 \log_{10}(H) \qquad \text{4-3}$$

$$c = 6.0560 - 0.0391 \cdot H - 0.0006 \cdot N \qquad \text{4-4}$$

with   H = dominant stand height (feet)
             N = stems per acre.

The parameter prediction method has been used by numerous authors, and it would be impossible to cover the many different approaches. A rather illustrative example is presented by Nagel and Biging (1995), who estimated b and c of the two-parameter Weibull function for several important tree species in Germany. They used the following regression equations:

$$b = \beta_0 + \beta_1 (Dg) \qquad \text{4-5}$$

$$c = \delta_0 + \delta_1 (Dg) + \delta_2 (Dmax) \qquad \text{4-6}$$

with   Dg = quadratic mean diameter [cm];
           Dmax = biggest observed diameter [cm].

The parameter values for Equations 4-5 and 4-6, involving five important tree species in Germany, are listed in Table 4-1.

Species		β_0	β_1	δ_0	δ_1	δ_2
Oak	(Quercus robur)	-1.937	1.082	4.669	0.366	-0.234
Beech	(Fagus sylvatica)	-4.282	1.132	4.518	0.317	-0.200
Spruce	(Picea abies)	-2.492	1.104	3.418	0.353	-0.192
Douglas Fir	(Pseudotsuga menziesii)	-0.621	1.060	4.380	0.236	-0.141
Scots Pine	(Pinus sylvestris)	-0.047	1.047	3.640	0.332	-0.180

Table 4-1. Coefficients for the Weibull parameter prediction Equations (4-5 and 4-6) for five important tree species in Germany.

A very simple approach for estimating the parameters of a diameter distribution was presented by Hui and Gadow (1996) for even-aged stands of *Cunninghamia lanceolata*. They used the logistic function (Eq. 4-7) as the diameter distribution model, which provided excellent fit to observed diameter frequencies. The parameters are directly derived from the diameters corresponding to the 50th and 90th percentiles. The percentile values are estimated from dominant stand height and quadratic mean diameter.

$$F(x) = \frac{1}{1 + e^{a-bx}}$$ 4-7

with b=2.1972/($X_{F=0.9}$- $X_{F=0.5}$) and a=-2.1972+$bX_{F=0.9}$

$X_{F=0.5} = 0.4043 \cdot H^{0.2762} Dg^{1.504 \cdot H^{-0.1403}}$ and $X_{F=0.9} = 1.2963 \cdot H^{0.1671} Dg^{0.7888 \cdot H^{+0.00668}}$

H=dominant stand height (m) and Dg=quadratic mean diameter (cm).

Fig. 4-3 presents two diameter distributions generated on the basis of the percentile estimates derived from the dominant stand height and the quadratic mean diameter.

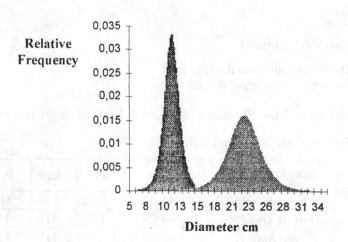

Figure 4-3. Two diameter distributions generated for Cunninghamia lanceolata stands using the logistic function. Left: Dg=12cm, H=10m; right: Dg=25cm, H=20m.

The method requires regular, unimodal conditions, which are found in well-managed timber plantations. It is not suitable for irregular stands. For even-aged forests exhibiting bimodal diameter distributions, more flexible models such as the *Charlier A* or the *Pearl-Reed* distribution functions may be used, although predictions of the function parameter values is often problematic (Álvarez González, 1997). Better results may be obtained by fitting two parameter-parsimonious unimodal distributions to observed frequencies with two distinct maxima. An example of this approach is the bimodal Beta distribution function applied by Maltamo et al. (1995) and Puumalainen (1996). The problem of modelling bimodal distributions was also addressed by Cao and Burkhart (1984).

Obviously, deriving distributions directly from average stand variables ensures compatibility. Joining two modelling levels in this way is one of the most effective strategies to generate detailed information, which is essential for estimating product yields, from a higher level, low resolution model.

Stand Table Projection

Stand table projection is one of the oldest techniques used to predict the future size-class distribution from a given initial one. The number of trees moving from one diameter class to the next within a discrete time step is known as *upgrowth*. Estimates of upgrowth are based on the *movement ratios* for each size-class, calculated as the mean increment divided by the class width.[1] This approach assumes that, after subtracting mortality, the trees are uniformly distributed in each diameter class. A useful approach to stand table projection involves a model that estimates diameter increment as a function of tree diameter.

---

[1] Example: 50% of the trees presently residing in diameter class i of width 4cm are assumed to move to class
     i+1 in a given time step, if $\Delta d_i = 2cm$.

Diameter Growth as a Function of Diameter

As instantaneous diameter growth dd/dt cannot be observed one has to rely on average periodic increment data for modelling the increase in stem diameter. The future diameter of a representative tree d_n may be predicted from the known initial diameter d_0 and the stand age $d_n = F(d_0, t) + \varepsilon$. A popular method for implementing this approach involves the use of the algebraic difference form of the so-called *Mitscherlich-equation*. The method was used by Saborowski (1982) and Lemm (1991) for predicting diameter growth of individual trees:

$$d_{2i} = d_{1i} \frac{1 - e^{-k(t_2 - t_0)}}{1 - e^{-k(t_1 - t_0)}}. \qquad 4\text{-}8$$

with t_1, t_2 = stand age at the beginning and end of the projection period;
d_{1i}, d_{2i} = breast height diameter of the i-th tree (cm) at age t_1 and t_2;
t_0 = age at which a tree reaches breast height (1,3 m);
k = empirical parameter.

Example: The diameter distribution of a 50-year old beech stand with a dominant height of 19m and normal (yield table) stocking was assessed during a stand inventory. According to Lemm (1991) the parameters k and t_0 of Eq. 4-8 may be estimated from dominant height as follows:

   $k = 0.003257 + 0.00016 \,(19) = 0,00356$

   $t_0 = 469 \, e^{-0,35379 \,(19)} = 0.56$.

The quadratic mean diameter of the stand is equal to 13cm (tree number 1), and the lower bound diameter equal to 9cm (tree number 2). The objective is to project the diameters of the two trees to age 55 years. Using Eq. 4-8 we obtain:

$$d_{21} = 13 \frac{1 - e^{-0,00356(55-0,56)}}{1 - e^{-0,00356(50-0,56)}} = 14.2 \text{cm} \quad \text{and} \quad d_{22} = 9 \frac{1 - e^{-0,00356(55-0,56)}}{1 - e^{-0,00356(50-0,56)}} = 9.8 \text{cm}.$$

The result reveals that throughout the projection period, the size ratios of the two trees did not change because 13/9 = 14.2/9.8. This result is not very realistic, as trees of different sizes growing within the same stand usually exhibit considerable differences in growth. Thus the assumed consistency of size ratios associated with the method represents a problem.[1]

At a given age, diameter growth in a stand usually increases with increasing diameter. This was already pointed out by Schwappach (1905), who found that in young and medium-aged stands in Northern Germany, the bigger oaks grew much faster than the smaller ones. However, these differences were not equally evident in the older oak stands (Fig. 4-4). Similar results were obtained in South African pine plantations, with the exception that diameter had no effect on growth in the very young, widely-planted stands (Gadow, 1984).

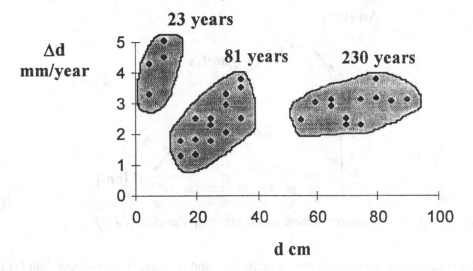

Figure 4-4. Relationship between breast height diameter (d) and diameter increment (Δd) in oak stands of different ages growing in Northern Germany (Bordesholm and Freienwalde, after Schwappach, 1905).

The examples demonstrate the potential for developing diameter growth models that estimate diameter increment (Δd) as a function of diameter (d) over a specified time period Δd=f(d) + ε. Vanclay (1994, p. 163 et sqq.) presented several models for predicting diameter increment from tree diameter. Two of

---

[1] The method is sometimes complemented using a stochastic component to overcome the problem (Sloboda, 1984; Gaffrey, 1996).

these, named after the authors Bertalanffy (1948) und Botkin (1993), were rearranged and the parameters selected such that they coincide approximately:

a) Bertalanffy: $\Delta d = 0.245 \cdot d^{0.44} - 0.0147 \cdot d$

$\qquad\qquad\qquad\qquad\qquad\qquad\qquad\qquad\qquad\qquad\qquad$ 4-9

b) Botkin: $\Delta d = \dfrac{d - d^2 \cdot \dfrac{137 + 50.9 \cdot d - 0.167 \cdot d^2}{611677}}{2.74 + 1.527 \cdot d - 0.00668 \cdot d^2}$ 4-10

Fig. 4-5 shows two similar growth curves for the Bertalanffy and Botkin functions, derived for sugar maple stands.

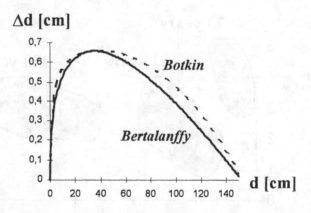

Figure 4-5. Two diameter growth functions after Vanclay (1994).

The relationship between diameter increment and diameter is age-dependent (*vide* Schwappach, 1905; Gadow, 1984). Thus in even-aged forests, when the stand age is known, Eq. 4-9 and 4-10 should be extended to include age.

Growth Modifiers

Diameter growth is closely related to crown dimensions, especially the crown surface area of a tree (Kramer, 1988). Dominant trees with large crowns grow better than suppressed trees with small crowns. A suitable measure of competition is the overtopping factor C66, as defined by Wensel et al. (1985).

The index *C66* is a measure of the relative position of a given subject tree within the population, based on the sum of the crown areas of all the other trees, measured at a given reference height. In order to calculate the index for a given subject tree, information about the total height and the height-to-crown base is required. The difference between total height and height-to-crown base gives the crown length, which is used to calculate the reference height (rh). The reference height for the subject tree is located at 66 percent of its crown length, measured from the top. The sum of the crown areas of all the other trees in the stand taken at or above this height is used to quantify the amount of competition assigned to the subject tree. Three possible situations may occur:

- The height-to-crown base of a competitor exceeds the reference height of the subject tree: the full crown area of the competitor is included,
– The height of the competitor is less than the reference height of the subject tree: the crown area of the competitor is not included,
- Otherwise, the crown area of the competitor at reference height is included.

For a given tree in a stand covering *a* square meters, the index C66 is calculated using Equation 4-11:

$$C66_i = \sum_j (KA_{66})_j / a \qquad \qquad 4\text{-}11$$

with KA = crown area of j competitor at reference height of subject tree [m²].

The value of the index C66 is defined by the reference height of the subject tree.

Example (provided by *J. Nagel*): Consider a beech stand with 4 trees, covering an area of 30m². The heights, diameters, crown dimensions and reference heights of the trees are shown in Fig. 4-6.

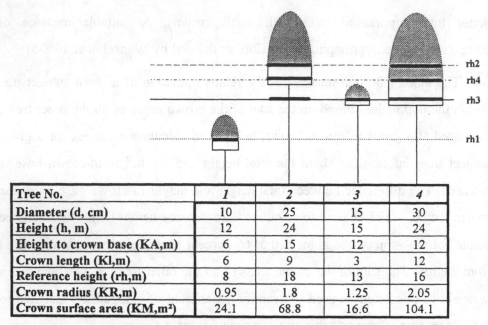

Tree No.	1	2	3	4
Diameter (d, cm)	10	25	15	30
Height (h, m)	12	24	15	24
Height to crown base (KA,m)	6	15	12	12
Crown length (Kl,m)	6	9	3	12
Reference height (rh,m)	8	18	13	16
Crown radius (KR,m)	0.95	1.8	1.25	2.05
Crown surface area (KM,m²)	24.1	68.8	16.6	104.1

Figure 4-6. Hypothetical beech stand with 4 trees, covering an area of 30 m².[1]

The C66 is obtained by summing the squared crown radii at or above the reference height of the subject tree, multiplying this value with π and dividing by the stand area. The following values are obtained for the four trees in Fig. 4-6:

Tree	C66
1	$\frac{\pi}{30} \cdot \left(0,95^2 + 1,8^2 + 1,25^2 + 2,05^2\right) = 1,04$
2	$\frac{\pi}{30} \cdot \left(0 + 1,8^2 + 0 + 1,77^2\right) = 0,67$
3	$\frac{\pi}{30} \cdot \left(0 + 1,8^2 + 1,25^2 + 2,05^2\right) = 0,94$
4	$\frac{\pi}{30} \cdot \left(0 + 1,8^2 + 0 + 2,05^2\right) = 0,78$

The index was used by Nagel (1994) to produce growth projections for different kinds of forests in North-Western Germany. Schübeler et al. (1995) used the index *C66* for calculating the height growth of a spruce tree of height *h* growing

---

[1] The variables *crown radius, crown surface area* and *crown radius at reference height* may be calculated as follows:

$$KR = \frac{0,843 + 0,11 \cdot d}{2} \quad ; KM = \frac{\pi \cdot KR}{6 \cdot Kl} \cdot \left[\left(4 \cdot Kl^2 + KR^2\right)^{\frac{3}{2}} - KR^3\right] ; \quad KR_{rh} = \sqrt{\frac{3 \cdot KR^2}{2 \cdot Kl} \cdot (h - rh)}$$

in a stand with dominant height H. The 5-year height growth Δh is estimated using the following potential-modifier function: $\Delta h = \Delta H(h/H)(\text{modifier})$ with modifier$=1.38-0.0645\ln(\text{age})-0.172(C66)$. The model was developed on the basis of height growth observations of 7635 trees growing in permanent plots in Germany.

Example: A spruce tree with a height of 15.0m is growing in a 70-year old stand with a dominant height of 17.7m. The value of the $C66$ is equal to 0.75, and the potential increase in dominant height during the next 5 years is 1.13m. Based on these data, the expected 5-year height growth of the tree is calculated as follows:

$$\text{mod ifier} = 1.38 - 0.064 \cdot \ln(70) - 0.172 \cdot 0.75 = 0.98$$

$$\Delta h = 1.13 \cdot \frac{15.0}{17.7} \cdot 0.98 = 0.94 \text{m}$$

Similar functions were developed for diameter growth, change in crown width and change in height-to-crown base.

The possibilities for developing growth modifiers such as the C66 are boundless. A good modifier is characterized by simultaneously accounting for the population attribute *stand density* and the tree attribute *relative position* within the population. As an example, consider the following growth modifier, which is applicable to the j'th tree growing in stand i:

$$m_j = 1 - R_i^{\{D_j/\overline{D}_i\}}$$

with $R_i = 1 - \sqrt{10000/N_i}/H_i$; D_j = diameter of tree j [cm]; $\overline{D}_i$ = mean diameter in stand i [cm] and H_i=dominant height in stand i [m].

Stand density increases with increasing value of R. A density-dependent negative effect on growth is associated with a high value of R. However, the growth of the largest trees in a stand is much less affected by density than that of the smaller ones. This phenomenon is accounted for by the exponent $D_j/\overline{D}_i$, which is a measure of the relative size of tree j within the stand. The index has some interpretable reference points. For example, $m_j=1$ at maximum possible density, i.e. if the average distance between the trees is zero. m_j assumes a value of zero if the average distance between the trees is equal to the dominant height. It will be negative when the average distance between the trees is greater than the dominant height. As an example, the value of m_j is calculated for four trees of different relative sizes growing in dense and in open stands.

	R_i	$D_i/\overline{D}_i$	m_i
large tree, dense stand	0.9	1.5	0.15
large tree, open stand	0.6	1.5	0.53
small tree, dense stand	0.9	0.6	0.06
small tree, open stand	0.6	0.6	0.26

A useful growth modifier for representative trees is one which simultaneously incorporates the density of a stand and the competitive position of the tree within the population. An example of a growth model resembling the potential-modifier approach is presented by Quicke et al. (1994) for even-aged stands of naturally regenerated Longleaf Pine (*Pinus palustris*) in the South-Eastern United States. All parameters of the model were estimated simultaneously, thereby avoiding the need to develop a separate function for estimating potential growth. The annual basal area increment of a tree is obtained from Eq. 4-12:

$$\Delta g = 11.52 \cdot e^{-0.0897 \cdot G^{0.5}} \cdot e^{-0.00397 \cdot BAL} \cdot e^{age \cdot \left(0.296\left\{1 - e^{-0.358 \cdot D}\right\} - 0.303\right)} \qquad 4\text{-}12$$

with Δg = annual basal area increment of a tree with diameter D [in²/year];
            G = stand basal area [ft²/acre];
         age = tree age [=7 plus the number of annual rings at 4 ft];
     BAL = sum of the basal areas of all trees larger than the reference tree;
       D = breast height diameter of reference tree [in].

The basal area increment of a tree is a function of stand density and age, and two additional components that are individual tree attributes. The BAL multiplier is a measure of the competitive position of a tree within the stand, while the diameter-age multiplier is a measure of historical vigor. The model performed well within the bounds of biologically reasonable outputs for any combination of values of the independent variables.

Change of Relative Basal Area

Future stand tables may be derived by means of a diameter distribution method or stand table projection based on diameter increment. The diameter distribution method ignores the availability of an observed initial stand table, and predicts or recovers future distribution parameters from projected stand variables such as the number of surviving trees per hectare, basal area per hectare and dominant height (Pienaar and Harrison, 1988). As mentioned before, among the problems associated with the stand table projection method are the *consistency of size ratios* and the *age dependence* of diameter growth equations used for estimating increment (Δd) as a function of diameter (d).

Clutter and Jones (1980) proposed a model that can be used to project an existing stand table based on the change in relative basal area. Relative basal area is defined as the ratio of the basal area of the i-th tree to the mean basal area. This ratio does not remain constant over time. It should increase in the dominant trees and decrease in the suppressed ones. The change of relative basal area may be described using the following equation:

$$\frac{g_{2i}}{\overline{g}_2} = \left[\frac{g_{1i}}{\overline{g}_1}\right]^{\left(\frac{t_2}{t_1}\right)^b} \qquad 4\text{-}13$$

with g_{1i}, g_{2i} = basal area of the ith surviving tree at stand age t_1 and t_2 (cm²),
$\overline{g}_1$, $\overline{g}_2$ = mean stand basal area at stand age t_1 and t_2 (cm²),
β = parameter to be estimated.

Equation (4-13) is independent of stand density and site quality. The important assumption is that the relative basal area changes with time. The model thus overcomes the problem associated with the *consistency of size ratios*. Equation (4-13) is compatible with the empirical observations of individual tree basal area development (Pienaar and Harrison 1988). Over a short period of time, the relative tree size ($g_i/\bar{g}$) remains constant. Over a longer period of time, the relative size of an individual survivor, and therefore the relative contribution of smaller than average-sized survivors to the total basal area, decreases, whereas the relative size of the largest trees increases. For the same period length, the change in relative size decreases, as age increases.

Example: The basal areas of two trees in a 50-year old hypothetical stand are: g_{11} = 132.7 cm² and g_{12} = 254.5 cm². The mean basal area at age 50 amounts to $\bar{g_1}$ = 201.1 cm². The mean basal area at age 55 is estimated at $\bar{g_2}$ = 283.5 cm² using a stand level growth model. After 5 years, with β=0.3, the following basal areas are predicted for the two trees:

$$g_{21} = 283.5\left[\frac{132.7}{201.1}\right]^{(55/50)^{0.3}} = 184.8 \text{ cm}^2$$

$$g_{22} = 283.5\left[\frac{254.5}{201.1}\right]^{(55/50)^{0.3}} = 361.2 \text{ cm}^2$$

The result shows that the model allows for a change in the size ratios. The size ratio between the two trees has increased as 361.2/184.8>254.5/132.7.

Forss et al. (1996) found ß = 0.7859 for *Acacia mangium* plantations in South Kalimantan, Indonesia. A projected stand table at age t_2, consistent with the projected per-hectare stand basal area, is obtained using the approach presented by Pienaar et al. (1990):

$$N_j g_{2j} = G_2 \frac{N_j\left(\dfrac{g_{1j}}{\bar{g_1}}\right)^a}{\displaystyle\sum_{j=1}^{k} N_j\left(\dfrac{g_{1j}}{\bar{g_1}}\right)^a} \qquad\qquad 4\text{-}14$$

where N_j= number of survivors in diameter class j (j=1,..,k) at age t_1;
  g_{1j}= basal area corresponding to midpoint of diameter class j at age t_1;
  g_{2j}= basal area corresponding to midpoint of diameter class j with N_j survivors at age t_2;
  g_1= average basal area of the N=ΣN$_j$ survivors at age t_1;

G_2= total basal area (m^2/ha) at age t_2;
$a = (t_2/t_1)^{0.78592}$.

The stand table projection procedure requires a prior estimate of future survival N_2 and of future per-hectare basal area G_2. The predicted total mortality is distributed over diameter classes in the initial stand table. This is accomplished by assuming that the probability of a tree in a given diameter class dying during the projection interval is inversely proportional to its relative size defined as $g_{j1} / \bar{g}_1$. Having identified the trees that will succumb during the projection period, the relative sizes of the diameter class midpoints are recalculated for the survivors only, and are then projected in compliance with Equation (4-14) to obtain future diameter class midpoints. The diameters of the survivors may then be distributed within diameter classes, by assuming that the trees are uniformly distributed around the projected diameter class midpoints.

A Worked Example

The procedure of projecting stand and tree parameters obtained from a stand inventory is illustrated for an *Acacia mangium* stand, which is 4 years of age at the time of inventory. The measured stand parameters are first projected up to the age of 5 years (Tab. 4-2).

	Current Stand	Projected Stand
Age (t, years)	4.0	5.0
Dominant height (m)	15.4	19.1
Stems/ha (N)	1574	1507
Basal area (G, m^2/ha)	17.1	22.9

Table 4-2. Current and projected stand variables using a stand growth model.

The next step involves projection of the initial stand table in such a manner that it remains consistent with projected survival and per-ha basal area. First, mortality (N_1-N_2) is calculated for each initial diameter class as shown in Tab. 4-3.

Diameter Class [cm]	N_{1j} Stems/ha (1)	G_{1j} [m²/ha] (2)	$\bar{g_1}/g_{1j}$ (3)	$(2)/\Sigma(2)$ (3)	$(1)/\Sigma(1)$ (4)	$(3)\times(4)$ $/\Sigma(3\times4)$ (5)	Mortal ity 67x(5)
2	62	0.019	34.50	0.420	0.039	0.383	26
3	77	0.054	15.33	0.186	0.049	0.212	14
4	62	0.078	8.63	0.105	0.039	0.096	6
5	31	0.061	5.52	0.067	0.020	0.031	2
6	77	0.218	3.83	0.047	0.049	0.053	4
7	46	0.177	2.82	0.034	0.029	0.023	2
8	15	0.075	2.16	0.026	0.010	0.006	0
9	46.	0.293	1.70	0.021	0.029	0.014	1
10	139	1.092	1.38	0.017	0.088	0.034	2
11	201	1.910	1.14	0.014	0.128	0.041	3
12	170	1.923	0.96	0.012	0.108	0.029	2
13	248	3.292	0.82	0.010	0.158	0.036	3
14	108	1.663	0.70	0.009	0.069	0.014	1
15	139	2.456	0.61	0.007	0.088	0.015	1
16	46	0.925	0.54	0.007	0.029	0.004	0
17	46	1.044	0.48	0.006	0.029	0.004	0
18	15	0.382	0.43	0.005	0.010	0.001	0
19	31	0.879	0.38	0.005	0.020	0.002	0
21	15	0.520	0.31	0.004	0.010	0.001	0
	1574	17.100	82.24	1.000	1.000	1.000	67

Table 4-3. Example showing the distribution of total mortality over diameter classes.

A new stand table is compiled based on the surviving trees in each diameter class, and the diameter class midpoints are projected to the age of 5 years (Tab. 4-4).

Diameter Class [cm]	N_j Stems/ha Survivors	G_{1j} [m²/ha]	$N_j\left(g_{1j}/\bar{g_1}\right)^a$ (1)	$(1)/\Sigma(1)$ (2)	$g_2=$ $G_2(2)/N_j$	d_2
2	36	0.011	0.51	0.000	2.1	1.6
3	63	0.045	2.34	0.001	5.4	2.6
4	56	0.070	4.14	0.003	10.7	3.7
5	29	0.057	3.64	0.002	18.3	4.8
6	73	0.206	14.17	0.009	28.2	6.0
7	44	0.169	12.33	0.008	40.8	7.2
8	15	0.075	5.78	0.004	56.0	8.4
9	45	0.286	22.96	0.015	74.2	9.7
10	137	1.076	89.84	0.057	95.4	11.0
11	198	1.882	162.95	0.104	119.7	12.3
12	168	1.900	170.12	0.108	147.3	13.7
13	245	3.252	300.24	0.191	178.3	15.1
14	107	1.647	156.46	0.099	212.7	16.5
15	138	2.439	237.85	0.151	250.7	17.9
16	46	0.925	92.47	0.059	292.4	19.3
17	46	1.044	106.84	0.068	337.9	20.7
18	15	0.382	39.92	0.025	387.2	22.2
19	31	0.879	93.86	0.060	440.5	23.7
21	15	0.520	57.65	0.037	559.1	26.7
	1507	16.900	1574.06	1.000		

Table 4-4. Calculations to project the stand table after removal of predicted mortality.

If the projection period is very short and the range of diameters large, then very small trees may show negative growth. Thus, the restriction had to be imposed that diameter growth may not be negative.

The method involves projecting the class boundaries. The future diameter classes in period 2 contain the same cohort of trees, excluding the ones that were removed as a result of estimated mortality. Note that the new diameter class midpoints in Table 4-4 are not integer, but real numbers. Small distortions may be caused by rounding so that midpoints are not always exactly median values. Class widths are unequal, ranging from 1 cm in the smallest class to 1.5 cm in the biggest one.

The method is an example of a compatible model in compliance with the telescope principle. Compatibility is enforced by the constraint imposed by Equation 4-14.

Transition Matrices

The most common stand table projection approach assumes that trees are uniformly distributed in each diameter class, and that a movement ratio can thus be calculated as the mean increment for the class divided by the class width. An example involving a stand with 21 trees distributed over five diameter classes is shown in Tab. 4-5. The distribution after one time step is the result of the different movement ratios in the diameter classes.

d_j	mr_{ij}	n_{1j}	n_{2j}
14	0.25	4	3
16	0.50	8	5
18	0.40	5	7
20	0.25	4	5
22	0.20	0	1
		21	21

Table 4-5. A hypothetical stand with 21 trees distributed over five diameter classes (mr_{ij}=movement ratio, i.e. proportion of trees moving from class i to class j; n_{1j}=initial number of trees; n_{2j}= number of trees after one time step).

The example was so chosen that classes do not contain the fractions of trees normally encountered after a projection. The movements of the trees are shown in Fig. 4-7.

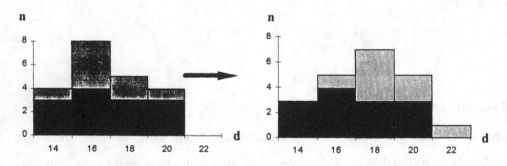

Figure 4-7. Graphical representation of the stand table projection based on the data presented in Tab. 4-5. Dark shading indicates the trees that remain in the classes, light shading the trees that move out (left) or in (right):

An alternative method for achieving a stand table projection as shown in Tab. 4-5 is the matrix approach. The matrix of movement ratios mr$_{ij}$ are multiplied with the initial state vector of tree diameters n$_{1j}$, which gives $n_{2j} = \sum_j mr_{ij} n_{1j}$. Using the numbers in Tab. 4-5 we obtain:

$$\begin{bmatrix} 0.75 & 0 & 0 & 0 & 0 \\ 0.25 & 0.50 & 0 & 0 & 0 \\ 0 & 0.50 & 0.60 & 0 & 0 \\ 0 & 0 & 0.40 & 0.75 & 0 \\ 0 & 0 & 0 & 0.25 & 0.80 \end{bmatrix} \times \begin{bmatrix} 4 \\ 8 \\ 5 \\ 4 \\ 0 \end{bmatrix} = \begin{bmatrix} 3 \\ 5 \\ 7 \\ 5 \\ 1 \end{bmatrix} \qquad 4\text{-}15$$

The columns of the matrix must sum to unity to ensure that the number of trees remaining, and those moving out, add up to the original class entry. This is not the case in column 5 in Eq. 4-15, simply because the matrix is incomplete, spanning only a fraction of the possible movement ratios. A plausible entry for the largest possible diameter would be 1 (in the last column and the last row of the matrix), indicating that there is no movement out of the class, due to stagnation in growth. Mortality may be modelled by reducing the numbers in the matrix so that the sum of the column entries is less than one. Thus, to introduce 2% mortality in the first diameter class, simply replace the value 0.75 in the first row, first column with 0.73.

The papers of Rudra (1968), Suzuki (1971), Moser (1974) and Sloboda (1976) were among the early methodical contributions covering aspects of diameter transition modelling. A very simple and appealing model for estimating movement ratios for spruce stands in Finland was developed by Kolström (1992; see also Pukkala and Kolström, 1988).

Kolström's Equation (4-16) estimates the probability that a randomly selected tree in the i-th 4cm-diameter class will move to the next higher class during the forthcoming 5 years. The class width and time step are fixed, thus ensuring that trees cannot move up more than one diameter class in a single time step.

$$b_i = e^{-2,1+0,86\ln(d_i)-0,55\ln(G)-0,0007G \cdot d_i} \qquad 4\text{-}16$$

with b_i = probability that a randomly selected tree in the i-th diameter class will move to the (i+1)-th class during the forthcoming 5 years; d_i = midpoint of diameter class i;

G   = stand basal area [m²/ha].

The movement ratio in class i is defined by the diameter and the stand density. It increases with increasing diameter and decreases with increasing basal area.

Example: Consider a diameter class with a midpoint of 20cm ($18 \leq d_i < 22$) comprising 100 trees in a spruce stand with a total basal area of 30m²/ha. Calculate the number of trees that will move into the 24-cm class during the next five years. Solution: $b_i = e^{-1.814} = 0.163$. Thus 16.3 trees will move into the class $22 \leq 24 < 26$cm, while 83.7 trees will remain in the class $18 \leq 20 < 22$cm, assuming zero mortality.

Figure 4-8. Eucalyptus grandis stand, South Africa.

Diameter-Height Relations

The height development of the size-class representative trees in an even-aged stand is rarely modelled directly. Height measurements are time-consuming and often inaccurate and initial height distributions are thus usually not available for projection. Instead, the heights are derived indirectly from the diameters, using a known or estimated relationship between diameters and heights (Van Laar and Akça, 1997, p. 149 *et sqq.*) and modelling the development of this relationship over time.

The relationship between diameters and heights may be described using a height regression or a bivariate diameter-height distribution. For modelling the *development* of the relationship, we may use a generalized height regression.

Generalized Diameter-Height Relations

A height regression may be derived separately for each stand, on the basis of pairs of diameter-height measurements obtained during a stand inventory. As this approach is costly, a practical alternative is to develop generalized height regressions which embody certain basic characteristics inherent in all individual height regressions (Wiedemann, 1935; Prodan, 1965; Wenk et al., 1990, p. 223; Kramer and Akça, 1987, p. 152; Hui and Gadow, 1993).

Generalized height regressions are an important element of size-class models. They are essential for estimating product yields for representative trees. At least 30 different functions have been used for describing the relationship between tree diameters and heights (Wenk et al., 1990, p.221). Two of these were found particularly suitable for deriving a generalized height regression. The first equation was derived on the basis of a function involving the logarithm of height as the dependent variable, and the reciprocal of diameter as the predictor variable (Eq. 4-17).

$$\ln(Y) = a_0 + a_1 \cdot \frac{1}{X} \qquad\qquad 4\text{-}17$$

with Y = some measure of tree dimension or stand production;
            X = predictor variable, e.g. stand age;
            a_0, a_1 = parameters.

This type of function was originally suggested in papers by MacKinney et al.
(1937) and Schumacher (1939). Numerous authors subsequently used the model,
which is known in the American forestry literature as the *Schumacher function*,
and in Central and Eastern Europe as the *Michailoff function* (after Michailoff,
1943). Eq. 4-17 was modified and used by several authors as a generalized height
regression:

$$h_i = 1.3 + (\overline{H} - 1.3)e^{a_1(1-\frac{\overline{D}}{d_i}) + a_2(\frac{1}{\overline{D}} - \frac{1}{d_i})} \qquad\qquad 4\text{-}18$$

with $\overline{H}$ = average stand height [m];
            $\overline{D}$ = average stand diameter [cm];
            h_i = height of i-th tree [m];
            d_i = breast height diameter of i-th tree [cm];
            a_1, a_2 = parameters to be estimated.

Equation 4-18 has only two parameters, which makes it an attractive modelling
tool. It was applied by Gaffrey (1988) and Sloboda et al. (1993) for estimating
the height/diameter relation in *Picea abies* forests. Nagel (1991) used the function
in stands of *Quercus rubra* and Goebel (1992) in *Fagus sylvatica*.

An alternative model was developed by Hui and Gadow (1993b) on the
basis of the allometric growth theory. However, instead of postulating a constant
ratio of the relative growth rates of the two parts of the organism (*vide* Kramer,
1988, p. 50), it was assumed that the ratio is influenced by a stand attribute such
as the dominant height:

$$\frac{\dfrac{1}{h_i}\dfrac{\partial h_i}{\partial t}}{\dfrac{1}{d_i}\dfrac{\partial d_{io}}{\partial t}} = aH^b \qquad\qquad 4\text{-}19$$

with
t = time;
H = dominant stand height [m];
h_i = height of i-th tree [m];
d_{i0} = diameter at the base of the i-th tree [cm];
a, b = parameters.

Integration of (4-19) gives (4-20).

$$\ln h_i = aH^b \ln d_{i0} + k \qquad\qquad\qquad 4\text{-}20$$

with k= constant of integration.

Equation (4-20) may be further developed so that eventually a generalized height regression is obtained. An example is Equation (4-21), which was used in stands of *Cunninghamia lanceolata* in Southern China.

$$h_i = 1.3 + a_1 H^{b_1} d_i^{\,a_2 H^{b_2}} \qquad\qquad\qquad 4\text{-}21$$

with
H = dominant stand height [m];
h_i = height of i-th tree [m];
d_i = breast height diameter of i-th tree [cm];
a_1, a_2, b_1, b_2 = parameters to be estimated.

Equation 4-21 proved to be very suitable when compared with other models, a particular advantage being its simplicity. Dominant height, representing the stand level, is the only additional variable required for estimating individual tree heights from diameters.

The following parameter values for Equation 4-21 were derived on the basis of 226 growth trials from the central region of the Jiangxi Province: a_1=0.162, b_1=0.880, a_2=1.198 and b_2=−0.212. The simulated development of the diameter-height regressions for *Cunninghamia lanceolata* is shown in Fig. 4-9. The diagram on the left in Fig. 4-9 shows the development for different ages, the one on the right for a given age and different site indices.

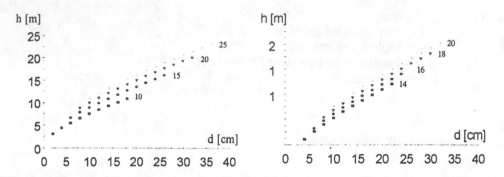

Figure 4-9. Simulated development of the diameter-height regression for a stand of Cunninghamia lanceolata. Left: site index 16, ages 10, 15, 20 and 25 years. Right: age 15, site indices 14, 16, 18 and 20.

Similar graphs showing the typical changes of the height regression lines are presented by a number of authors (Prodan, 1965, p. 178; Gadow, 1987, p. 81; Kramer, 1988, p. 83; Wenk et al., 1990, p. 223; Nagel, 1991).

Bivariate Diameter-Height Distributions

Size-class distribution information is important because it affects the type and timing of management strategies and treatments applied, and it influences the growth potential and hence, the current and future economic value of the forest stand (Knoebel and Burkhart, 1991). A height-diameter regression is often not very appropriate because it produces only one height for each given diameter. A more realistic approach assumes that there is a distribution of heights corresponding to a given diameter class. Hence, there has been some interest in the successful application of a bivariate statistical distribution for describing diameter-height frequency data.

The bivariate extension of the univariate S_B distribution (Johnson, 1949a), which is known as the S_{BB} distribution (Johnson, 1949b), may provide useful information by describing even-aged forest height-diameter data (Hafley and Schreuder, 1977).

The S_{BB} is a *bivariate* distribution for which both marginals[1] are S_B distributions. It is the bivariate distribution of y_1 and y_2 when the standard normal variates z_1 and z_2 are defined as:

$$z_1 = \gamma_1 + \delta_1 \ln\left\{y_1 / (1 - y_1)\right\} \quad \text{and} \quad z_2 = \gamma_2 + \delta_2 \ln\left\{y_2 / (1 - y_2)\right\} \qquad \text{4-22}$$

where z_1 and z_2 have the joint bivariate distribution with correlation ρ, namely,

$$p(z_1, z_2; \rho) = \left[2\pi\sqrt{1 - \rho^2}\right]^{-1} \exp\left\{-(1/2)(1 - \rho^2)^{-1}\left(z_1^2 - 2\rho z_1 z_2 + z_2^2\right)\right\}$$

Here, $y_1 = (d - \xi_1) / \lambda_1$ and $y_2 = (h - \xi_2) / \lambda_2$ and ξ_1 and ξ_2 represent the smallest values, and λ_1 and λ_2 the range of diameters (d) and heights (h), respectively, in the population. Many of the properties of the S_{BB} can be obtained directly from the literature of the bivariate normal distribution, because of the relation between them.

The correlation coefficient ρ is obtained from

$$\rho = \sum_{j=1}^{n} z_{1j} z_{2j} / n \qquad \text{4-23}$$

where, $z_{ij} = \gamma_i + \delta_i \ln\left\{y_{ij} / (1 - y_{ij})\right\}$; $y_{ij} = (X_{ij} - \xi_i) / \lambda_i$ with $X_{1j} = d_j$ and $X_{2j} = h_j$; n = number of diameter-height pairs of data; i = 1, 2; j = 1, ···, n. The parameters γ and δ are estimated as follows:

---

[1]Relative to the joint density $f_{X,Y}(x,y)$, the densities of X and Y, $f_X(x)$ and $f_Y(y)$, are called the marginal densities. From a practical perspective, the marginal distributions describe the individual behavior of the measurements X and Y, but they often lack, even when taken together, information regarding the joint behavior of the measurements. Although the marginal densities can be recovered from the joint density, in general, the joint density can not be constructed from a knowledge of the marginal densities. In the discrete case, the marginal density $f_X(x)$ is found by holding x fixed and summing over y, whereas in the continuous case, it is found by holding x fixed and integrating over y. An analogous comment applies to finding $f_Y(y)$ (see Dougherty, 1990, p. 208-9 & 215).

$$\gamma_i = (-f_i)/s_i$$

$$\delta_i = 1/s_i$$

where, $f_i = (\sum_{j=1}^{n} f_{ij})/n$; $s_i^2 = 1/n \sum_{j=1}^{n} (f_{ij}-f_i)^2$; $f_{ij} = \ln\{y_{ij}/(1-y_{ij})\}$; n and y_{ij} is as above; i = 1, 2; j = 1,···, n.

One of the properties of interest is the regression relation between y_2 and y_1. In general, the usual mean regression is complicated. However, the median regression takes a much simpler form, namely:

$$y_2 = \theta y_1^\phi \{(1-y_1)^\phi + \theta y_1^\phi\}^{-1} \qquad\qquad 4-24$$

where

$$\theta = \exp\{(\rho\gamma_1 - \gamma_2)/\delta_2\} \text{ and } \phi = \rho\delta_1/\delta_2 \qquad\qquad 4-25$$

Here, $\phi > 0$ since it is assumed that $\rho > 0$. The shape of the regression curve is influenced by ϕ, and the slope depends on the range of values of ρ. For $0 < \phi < 1$, the first derivative of y_2 with respect to y_1 does not exist at the extremes, $y_1 = 0$ and $y_1 = 1$. For data sets of diameter and height, the constraint $\phi > 1$ should be imposed on parameter estimation. The regression is linear if $\rho\delta_1 = \delta_2$ and $\rho\gamma_1 = \gamma_2$. An important aspect of the equations is that ϕ and θ are functions of the two marginal distribution parameters. Rewritten in terms of diameter and height, Equation (4-24) can be expressed as

$$(h - \xi_2)/\lambda_2 = \theta \left[\{(\xi_1 + \lambda_1 - d)/(d - \xi_1)\}^\phi + \theta \right]^{-1} \qquad\qquad 4-26$$

A second property of interest is the fact that the conditional distribution of y_2 given y_1 also follows the S_B distribution with parameters γ' and δ' where

$$\gamma' = [\gamma_2 - \rho z_1](1-\rho^2)^{-1/2} \text{ and } \delta' = \delta_2(1-\rho^2)^{-1/2}.$$

This conditional distribution can be bimodal[1]. Tewari and Gadow (1995) fitted the distribution to the diameters and heights of a 10-year old *Acacia tortilis* plantation consisting of 251 trees, located in the hot arid region of Rajasthan in India. Diameters were measured to the nearest 0.1 cm and heights to the nearest 5 cm. The statistics for the diameter-height data of the *Acacia tortilis* stand are given in Table 4-6.

Distribution parameters	Diameters (cm)	Heights (m)
Mean	7.11	4.47
Variance	4.78	0.55
Index of Skewness	0.3207	-0.3263
Index of Kurtosis	3.0680	3.1016

Table 4-6. Distribution parameters for diameters and heights of Acacia tortilis.

The diameter-height data pairs were used to fit the S_{BB} distribution using the computer program 'GETIT.PAS' developed by W.L. Hafley. First, the data are used to estimate the parameters of the marginal S_B distribution. The results are presented in Table 4-7.

S_B Parameters	Diameter	Height
Gamma (γ)	0.9686	-0.5729
Delta (δ)	1.8809	1.6460
Mu (μ)	-0.5150	0.3480
Sigma (σ)	0.5317	0.6075

Table 4-7. Parameters of the S_B distributions fitted to observed diameters and heights in an Acacia tortilis stand.

---

[1]The extraction of information about one random variable, when given information about the behavior of another, leads to the problem of conditioning. If X and Y are discrete and posess the joint density F(x, y), then conditional probabilities regarding Y take the form of conditional probabilities concerning the relevant events. The probability that Y = y, given that X = x, is

$$P(Y = y \mid X = x) = P(Y = y, X = x)/P(X = x) = f(x, y)/f_x(x)$$

Intuitively, the observation X = x is recorded, and one is left with probabilistic knowledge concerning the random variable Y. Since Y is discrete, only the conditional probabilities of its possible outcomes need to be specified. In the case of jointly distributed continuous random variables, as long as $f_x(x) \neq 0$, $f(x, y)/f_x(x)$ is still defined. This expression is taken as the general definition of the conditional density. The random variable associated with the conditional density is known as the conditional random variable of Y given x, and is denoted by Y|x (see Dougherty, 1990, p. 249).

The parameters γ_i and δ_i ($i = 1, 2$) are used to transform the pairs of observations to standard normal deviates which, in turn, are used to calculate the marginal S_B densities (Table 4-8). Thus, the marginal densities are not recovered from the bivariate S_{BB} joint distribution, but obtained by fitting the univariate S_B distribution model.

Diameter				Height			
Class Bounds	Observed Frequency	Predicted Frequency	Residual	Class Bounds	Observed Frequency	Predicted Frequency	Resid.
0.0 - 2.0	2.0	0.3	-1.7	1.3 - 2.0	0.0	0.0	-0.0
2.0 - 4.0	15.0	17.4	2.4	2.0 - 4.0	76.0	68.1	-7.9
4.0 - 6.0	64.0	65.8	1.8	4.0 - 6.0	173.0	181.0	8.0
6.0 - 8.0	90.0	84.0	-6.0	6.0 - 6.8	2.0	1.9	-0.1
8.0 - 10.0	59.0	56.7	-2.3				
10.0 - 12.0	16.0	22.1	6.1				
12.0 - 14.0	5.0	4.4	-0.6				
14.0 - 18.6	0.0	0.3	-0.3				

Table 4-8. Marginal S_B distributions separately for diameters and heights of Acacia tortilis.

The next step involves calculating the correlation coefficient ρ using the standard normal deviates in Equation (4-23). Further, the correlation coefficient ρ and marginal S_B distribution parameters are substituted in Equation (4-25) to obtain the S_{BB} parameters (Table 4-9).

S_{BB} Parameters	Values
Rho (ρ)	0.5428
Theta (θ)	1.9492
Phi (ϕ)	0.6200

Table 4-9. Parameters of the bivariate S_{BB} distribution.

Finally, the correlation coefficient ρ and the standard normal deviates z_{ij} are used to obtain the S_{BB} joint density distribution (Table 4-10).

Diameter Class (cm)	Height Class (m)					
	2.0 - 4.0		4.0 - 6.0		6.0 - 6.8	
	obs.	pred.	obs.	pred.	obs.	pred.
0.0 - 2.0	**0.0080**	*0.0011*	0.0000	*0.0001*	0.0000	*0.0000*
2.0 - 4.0	**0.0438**	*0.0475*	0.0159	*0.0218*	0.0000	*0.0000*
4.0 - 6.0	**0.1155**	*0.1144*	0.1394	*0.1478*	0.0000	*0.0000*
6.0 - 8.0	**0.0996**	*0.0793*	0.2590	*0.2547*	0.0000	*0.0007*
8.0 - 10.0	**0.0319**	*0.0249*	0.1992	*0.1988*	0.0040	*0.0021*
10.0 - 12.0	0.0000	*0.0037*	0.0598	*0.0814*	0.0040	*0.0028*
12.0 - 14.0	**0.0040**	*0.0002*	0.0159	*0.0156*	0.0000	*0.0016*

Table 4-10. Observed and predicted S_{BB} joint distributions for the height and diameter classes.

The predicted and observed frequencies in Table 4-10 are almost identical in the main part of the data range. Minor differences in the tail regions may be due to the limited sample size. The corresponding graph is shown in Figure 4-10.

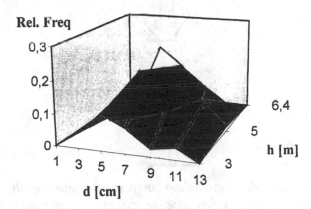

Figure 4-10. Predicted bivariate frequencies for three height and 7 diameter classes.

The primary advantage of the S_{BB} approach is that it permits a more realistic description of the height-diameter relation than the traditional regression approach. However, it is not very likely that sufficient diameter-height data are always available. Thus, a problem which would have to be adressed in the future, would be the development of *generalized* bivariate distributions involving predictions of the parameter values for different forest types and management regimes.

Estimating Product Yields

The key to successful timber management is a proper understanding of the variables that influence tree growth and stand development. Growth and yield modelling is the essence of the timber business. Foresters need to be able to anticipate the consequences of a thinning in terms of the yield potential. However, total timber volume is usually not a very appropriate measure for describing yields. The value of timber is defined by the dimensions and quality attributes of the saleable logs (Fig. 4-11). Thus, models are needed for predicting product yields in standing trees, for a variety of environmental and treatment conditions, enabling foresters to take management and investment decisions at stand level.

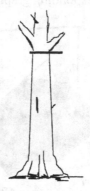

Figure 4-11. Product yield modelling involves estimating the volumes of different products with given size specifications and quality attributes. Estimating the height to the base of the crown is an important modelling issue in hardwood forests.

Estimating timber products is a central issue in growth and yield research, and one of the basic requirements is the ability to predict stem form in order to be able to simulate the assortment yields in a standing tree. However, besides being able to describe the shape of a tree, it is also essential to estimate the occurrence of quality attributes at different heights of the trunk, which is especially important in oak, beech and pine forests of high value. Thus, product yield modelling involves estimating the volumes of different products with given *size specifications* and *quality attributes*.

Volume Ratio Methods

A popular approach for estimating products with given size specifications is the volume ratio method. The equation used by Burkhart (1977) can be applied to calculate the volume to a given thin-end diameter m. Trincado and Gadow (1996) proposed the following modification for hardwoods:

$$R_d = \frac{V_d}{Vka} = 1 + b_0 \left[\frac{m - Dkra}{d - Dkra} \right]^{b_1} \qquad m \geq Dkra\,;\ d > Dkra \qquad \text{4-27}$$

with R_d = volume ratio to a given thin-end diameter m (cm);
V_d = volume (m³) over bark to a given thin-end diameter m (cm)
and with a given diameter at the base of the crown $Dkra$ (cm);
Vka = over bark volume from the base of the tree to the base of the crown (m³);
d = diameter at breast height 1,3 m (cm);
b_0, b_1 = parameters.

The modification ensures compatibility. The volume up to the crown base is obtained when the thin end diameter m is equal to the diameter at crown base Dka.

Another well-known volume ratio model is the one published by Cao et al. (1984). This function may be adjusted for hardwood trees to take the following form:

$$R_h = \frac{V_h}{Vka} = 1 + b_0 \left[1 - \frac{h}{Hka} \right]^{b_1} \qquad h \leq Hka \qquad \text{4-28}$$

with R_h= volume ratio at a given tree height h (m);
V_h= volume (m³) over bark up to a given tree height h (m)
and with a given height to base of crown Hka (m).

Again, the modification was done in such a manner that the volume to the crown base is obtained when the height to base of crown Hka and the tree height h are identical. The model proposed by Honer (1965) may be used to estimate the total over bark volume from the tree base to the crown base, Vka, using breast height diameter d:

$$Vka = b_0 + b_1 \cdot d^2 \cdot Hka \qquad \text{4-29}$$

The Equations 4-27, 4-28 and 4-29 were fitted using the data of 542 beech *(Fagus sylvatica)* trees grown in various parts of Northern Germany. The estimated parameter values are presented in Tab. 4-11.

Function	Parameter Values		Statistics	
	b_0	b_1	R^2	Error
(4-27)	-0.8941	1.5399	0.8677	0.0870
(4-28)	-0.9499	1.3202	0.9907	0.0231
(4-29)	0.018787	0.00005235	0.9860	0.1170

Table 4-11. Parameter estimates for Equations (4-27), (4-28) and (4-29) based on 542 beech trees.

Based on Equations (4-27) and (4-28), a new model may be derived, which can be used to calculate the diameter at a given tree height, or the height corresponding to a given stem diameter:

$$d_h = Dka + \left[\frac{(R_h - 1)}{b_0 (d - Dka)^{-b_1}} \right]^{\frac{1}{b_1}} \qquad 4\text{-}30$$

with d_h = stem diameter [cm] at a given tree height h, d=tree breast height diameter [cm] and R_h calculated using Equation 4-28.

Example: Use the parameter values in Tab. 4-11 to estimate the diameter and the stem volume of a standing beech tree with d = 60.9 cm, *Hka* = 20.2 m and *Dka* = 42.3 cm at a height of 16m. Solution: d_{16}=47.33 cm und V_{16}= 3.470 m³.

Modelling Stem Profiles: Form Quotients, Splines and Polynomials

Estimates of assortment distributions in standing trees are primarily based on descriptions of stem form. The *form factor*, for example, the ratio of the observed stem volume to an ideal cylinder, is a very simple measure of stem form. A more detailed description is provided by a *form quotient*, which represents the ratio of the tree diameters at two different heights (Prodan, 1965; Akça et al. 1994). The logical extension of this concept is a vector of form quotients for describing an entire stem profile. Such taper series have been in use for many years.[1]

An interesting application of the taper series concept using a linear model was presented by Sloboda (1984) and Gaffrey (1988). In this application, a series of linear relationships were established between the breast height diameter and the diameter at different relative tree heights. The parameters of the various relations were then predicted using a 5th-degree polynomial.

Another popular method for describing stem profiles is based on applications of spline functions. The method requires diameter measurements taken at n heights $[(d_i, h_i) i=1,..,n]$ of a tree. The spline function describes a stem profile through a series of polynomials which are linked in a particular way. In order to obtain a complete description of the profile, all coordinates in the interval $[h_i, h_{i+1}]$ must be determined by interpolation. The most common application is a cubic spline function representing n-1 third-degree polynomials

$$f_i(x) = \alpha_i + \beta_i(x - h_i) + \gamma_i(x - h_i)^2 + \delta_i(x - h_i)^3, \ x \in [h_i, h_{i+1}]$$ 4-31

which are smoothly linked at the nodes using their second derivatives. Given *n* nodes this approach requires *n-1* sets of coefficients, i.e. a total of 3(n-1) coefficients (Späth, 1973, p. 27f.; Saborowski, 1982, p. 7f.; Gaffrey, 1988, p. 51f.).

Series of form quotients, high-degree polynomials and spline functions are capable of describing, with high precision, a *particular* stem profile. However, this ability is achieved at the cost of a great number of parameters, and what appears to be an advantage at first glance, turns out to be a serious handicap. It is impossible to use taper series and spline functions for developing generalized

---

[1] Vectors of form quotients are known as taper series or *Ausbauchungsreihen* (Grundner u. Schwappach, 1942; Schober, 1952; Gadow et al., 1995).

stem profile models, which describe stem form as a function of environment and silviculture.

Parameter-parsimonious Stem Profile Functions

The stem represents an assortment of saleable timber products and, for comparing different cross-cutting options, its shape needs to be known. A simple model of stem geometry is the *form factor*, i.e. the stem volume expressed as a proportion of the volume of a theoretical cylinder defined by the tree's height and diameter at 1.3m. Form factors can be used to estimate total tree volume, but they cannot describe a stem profile and are thus not suited for product modelling. On the other hand, rather a lot is known about the general relationship between form factor, stand density and tree age.

Stem forms can be described in great detail using spline techniques or high-degree polynomials. Such very accurate profile descriptions are required in a sawmill where optimum cutting options are sought for individual logs. As previously pointed out, however, too much precision can be a serious disadvantage. The simpler model is often the more useful one, because it permits estimates of tree form in response to silviculture. However, when the model is too simple, as in the case of the *form factor*, it does not produce the desired minimum information for estimating timber product yields.

The detailed models cannot be used to make general statements about the way in which silviculture affects stem form, while the simple ones are not capable of producing profile descriptions that are sufficiently accurate. A suitable compromise is a taper function with a limited number of parameters.[1] The

---

[1] See for example, Demaerschalk (1973); Clutter (1980); Reed u. Green (1984); Brink u. Gadow (1986); Kozak (1988); LeMay et al. (1993); Nagashima u. Kawata (1994).

following *modified Brink function*[1] is an example of a parameter-parsimonious stem taper model (Riemer et al., 1995):

$$r(h') = u + v \cdot e^{-ph'} - w \cdot e^{qh'} \qquad\qquad 4\text{-}32$$

with $u = \dfrac{i}{1 - e^{q(1,3-h)}} + (r_{13} - i)\left(1 - \dfrac{1}{1 - e^{p(1,3-h)}}\right)$, $v = \dfrac{(r_{13} - i) \cdot e^{p \cdot 1,3}}{1 - e^{p(1,3-h)}}$ and $w = \dfrac{i \cdot e^{-qh}}{1 - e^{q(1,3-h)}}$

and $r(h')$ = tree radius (cm) at height h' (m);
       h = total tree height (m);
       r_{13} = tree radius at breast height (cm);
       i = parameter (common asymptote);
       p = parameter (describing lower part of stem);
       q = parameter (describing upper part of stem*).*

Equation 4-32 complies with the constraints that r(h')=0 when h'=h, and that r(h')=r_{13} when h'=1.3.

A number of successful applications of the modified Brink function involving trees of different species and ages have been published (Riemer et al., 1995; Trincado, 1996; Trincado and Gadow, 1996; Hui and Gadow, 1997; Van Laar and Akça, 1997, p. 182). Steingaß (1995) fitted the function to 2882 Douglas fir trees. The fitted stem profiles of the 5, 50 and 95-percentile trees are shown in Fig. 4-12.

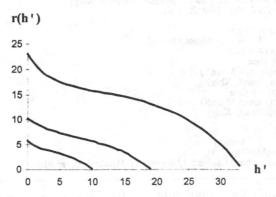

Figure 4-12. Fitted stem profiles of the trees corresponding to the 5, 50 and 95-percentiles in a sample of 2882 Douglas fir trees.

---

[1] Named after C. Brink, who had a major share in developing the model (Brink and Gadow, 1986).

The parameters of the trees corresponding to the profiles in Fig. 4-12 are listed in Tab. 4-12.

Tree	d [cm]	h [m]	i	p	q	Age
95-percentile No. 7120225	41.4	33.5	16.45	0.3259	0.1023	86
Representative No. 3111077	17.8	19.1	7.39	0.3096	0.1281	46
5-percentile No. 3223316	9.2	10.2	4.24	0.9999	0.1966	38

Table 4-12. Parameters corresponding to the three stem profiles shown in Fig. 4-12.

Example: Calculate the diameter at a height of 15m for a 95-year old spruce tree with a breast height diameter of 41cm and a height of 35m. Use the modified Brink function with parameters i=17.5, p=1.0 and q=0.03. Solution: 2*12.41=24.82 cm. Calculate for the same spruce tree the height which corresponds to a diameter of 8 cm. Solution: 29.76 m (this value is obtained iteratively using the root function in the program *Brink2*).

```
Program Brink2;
{calculates h for a given radius}
Const i = 17.5; p = 1.0; q = 0.03; d = 41; h = 35;
Var   Radius, Height: real;

Function F(ht:real):real;
Var u, v, w: real;
Begin
  u:=i/(1-exp(q*(1.3-h)))+(d/2-i)*(1-1/(1-exp(p*(1.3-h))));
  v:=(d/2-i)*exp(p*1.3)/(1-exp(p*(1.3-h)));
  w:=i*exp(-q*h)/(1-exp(q*(1.3-h)));
  F:=ln((u+v*exp(-p*ht)-Radius)/w)/q-ht;
End; {Radius}

Function Root(Xsmall, Xbig: real):real;
{Recursive Funktion delivering root F between Xsmall & Xbig}
Const Epsilon = 1E-8; Var  MeanX: real;
Begin
  MeanX:=(Xsmall+Xbig)/2;
  if Xbig-Xsmall < Epsilon then Root:=MeanX
  else if F(MeanX) > 0 then
    if F(Xbig) > 0 then
      Root:=Root(Xsmall,MeanX) else
      Root:=Root(MeanX, Xbig)
    else if F(Xbig) < 0 then
      Root:=Root(Xsmall,MeanX)  else
      Root:=Root(MeanX, Xbig)
End; {Root}

BEGIN                        {main}
  Radius:=4;  Height:=Root(0,h);
  writeln('height at radius ',Radius:5:1,' is equal to',Height:5:2);  readln;
END.
```

Program 4-1. Code for calculating the tree height corresponding to a given radius.

The stem volume of a tree $V(h_1,h_2)$ between two heights h_1 and h_2 can be calculated analytically, using the integral of the squared function

$$V(h_1,h_2) = F(h_2) - F(h_1) \quad \text{with} \quad F(h') = \int f(h') \, dh \text{ and } f(h') = \pi \cdot (r(h'))^2 \quad 4\text{-}33$$

For the modified Brink-function we obtain:

$$F(h) = \pi \int u^2 + v^2 e^{-2ph'} + w^2 e^{2qh'} dh' + 2\pi \int uve^{-ph'} - uwe^{qh'} - vwe^{(q-p)h'} dh'$$

$$= \pi \left(u^2 h' - \frac{v^2}{2p} e^{-2ph'} + \frac{w^2}{2q} e^{2qh'} - \frac{2uv}{p} e^{-ph'} - \frac{2uw}{q} e^{qh'} + \frac{2vw}{p-q} e^{(q-p)h'} \right) \qquad 4\text{-}34$$

Example: Calculate the volume of the bottom 3m log for the spruce tree used in the previous example. The height of the stump is 0.25 m. Solution: Using the program _volume_, we obtain 0.3855 m³ over bark between h'=0.25 m and h'=3.25 m.

```
Program Volume;
Const i=17.5; p=1.0; q=0.03; DBH=41.0; h=35.0; from=0.25; until=3.25;
Var Vol: real;

Function Cumvol(DBH,h,i,p,q,Height: Real): real;
Const bh=1.3;              {breast height}
Var  s,t,u,v,w: real;
Begin
  t := bh-h;
  s := 1/(1-exp(p*t)); t := 1/(1-exp(q*t)); w:= 0.5*DBH-i;
  u := i * t + w * (1-s); v:=w*exp(p*bh)*s;
  w := i*exp(-q*h)*t; s:=exp(-p*Height); t := exp(q*Height);
  Cumvol:=pi/10000*(sqr(u)*Height+2.0*(v*w*s*t/(p-q)
    -u*w*t/q-u*v*s/p)+0.5*(sqr(w)*sqr(t)/q-sqr(v)*sqr(s)/p));
End; {Cumvol}

BEGIN {main}
  vol:= cumvol(DBH, H, i, p, q, until)
    -cumvol(DBH, H, i, p, q, from);
  writeln('Volume:', vol:7:4, ' [m³]');
  readln;
END.
```

Program 4-2. Code for calculating the volume for a given stem section.

An advantage of the modified Brink function is the fact that log volumes can be calculated directly using analytical integration. The model fits observed trees of different species and sizes very well, including the critical basal sweep. The parameters can be associated with certain parts of a tree and thus have some biological significance. A further advantage of the modified Brink function is the fact that it can be used for developing a generalized stem profile function.

Generalized Stem Profile Functions

A generalized taper function is used to predict tree profiles for a variety of environmental conditions using as few additional variables as possible. It is convenient to use a parameter-parsimonious function such as Equation 4-32. Gadow et al. (1996) found that the parameter i was related to breast height diameter and the parameter q to the height of spruce trees, while Steingaß (1996) was not able to establish a relationship between the parameter p and the height of Douglas fir trees. These findings were complemented by a study involving a *Cunninghamia lanceolata* data set from the Hunan Province in China, with a total of 1950 measurements of 223 trees, which were grown in 63 plots of 600m² each, subjected to different silvicultural treatments.

Diameter measurements were taken at the standard heights of 0.0, 1.3, 3.6, 5.6, 7.6m etc. Above 3.6m, the standard measurement intervals were 2m. In addition, some trees were measured at 0.5m to obtain data for modelling the butt sweep. Missing values at 0.5m were obtained either by interpolation between 0 and 1.3m or by using a linear model of the form $D_{0.5}=a+bD_{1.3}$, if $D_{0.5}-D_{1.3} > 3$cm. Important tree and stand parameters are listed in Tab. 4-13.

Value	h [m]	d [cm]	Dg	N	Age
Smallest	5.8	5.5	11.5	750	18
Average	15.2	19.1	17.9	1667	25
Biggest	22.8	29.5	23.3	4500	35

Table 4-13. *Tree and stand parameters used for establishing a generalized taper function (h=height of sample tree [m]; d=breast height diameter of sample tree [cm]; Dg= quadratic mean diameter of plot from which the sample tree was taken in [cm]; N=stems per ha of plot from which the sample tree was taken).*

The following relations could be established for predicting the parameters i, p and q from breast height diameter (d) and height (h) of a sample tree and the quadratic mean diameter (Dg) of the stand in which the tree is found:

$$i = k_1 d^{k_2}$$ 4-35

$$p = e^{\frac{k_3}{Dg}}$$ 4-36

$$q = k_5 e^{\frac{k_6}{Dg}} h^{k_4}$$ 4-37

with d=breast height diameter of tree [cm]; h=total height of tree [m];
     Dg=quadratic mean diameter of stand; the k's are parameters.

The quadratic mean diameter (Dg) is an important stand variable, which is usually known. It is therefore convenient to consider Dg when developing the form of a generalized taper model. A generalized taper function is obtained by substituting i, p and q in (4-32) with the right-hand sides of (4-35), (4-36) and (4-37). The stem profiles of trees with known breast height diameters and heights, growing in a stand with a known quadratic mean diameter, can now be estimated.

The following parameter estimates were obtained:

$k_1=0.40788$; $k_2=1.03702$; $k_3=1.79624$; $k_5=10.43850$; $k_6=-1.41743$; $k_4=-1.50117$

The coefficient of determination was 0.968 with n=2158, the error mean square 0.397. An F-value of 2.521 was obtained for assessing bias[1]. There is no reason to assume bias, as $F=2.521 < F_{0.05(2/2156)}=3$. An example of the generalized taper function superimposed on the observed stem radii of a tree with Dg=15.2cm, D=15.6cm and H=15.2m is shown in Fig. 4-13.

---

[1] See Chapters 3 and 6 with more detail about this test.

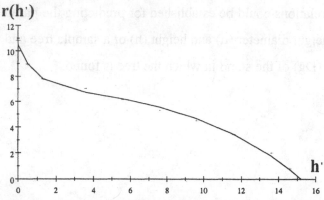

Figure 4-13. Example of the generalized taper function superimposed on the observed radii of a tree with d=15.6cm and h=15.2m. The quadratic mean diameter (Dg) of the stand is equal to 15.2cm.

An obvious standard for comparison involving the modified Brink function is provided by the tree volume function for *Cunninghamia lanceolata*, used in China:

$$vol = 0.000058777 \cdot d^{1.9699831} h^{0.89646157} \qquad\qquad 4\text{-}38$$

with d=breast height diameter (cm), h=tree height (m) and vol=individual tree volume (m³).

The total under bark stem volume was calculated for each of the 223 sample trees using Equation 4-38. The generalized Brink function with parameter values estimated by Eqs. 4-35, 4-36 and 4-37 was then used to calculate the volumes for the same trees.

Figure 4-14 presents the results of the comparison. Obviously, the agreement is very good. The generalized taper function can also be used for simple volume calculations, and this fact should be a convincing argument when recommending the practical use of a more sophisticated modelling tool.

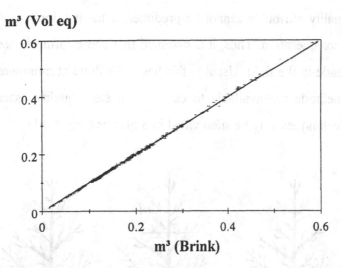

Figure 4-14. Comparison of the under bark stem volume estimates of the 223 sample trees using the Brink function (Brink) and the volume equation (Vol. Eq.).

Obviously, estimating total stem volumes is the least important of the potential applications of a generalized taper function. A major application is the calculation of log assortments, for given log specifications, in forests in which the diameter distribution is known. The ability to calculate product volumes of standing trees is an essential basis for estimating potential future yields and thus indispensable for evaluating silvicultural alternatives.

Stem Quality Assessment and Prediction

The quality assessment of standing trees, a prerequisite for silvicultural decision-making in stands including trees of high value, is an important and difficult task. A major problem is the uncertainty associated with the inner quality attributes of trees, which are not recognizable from the outside. Another predicament is the deficiency of inventory and forecasting methods. Existing knowledge in the field of statistical sampling theory has not yet been applied to stem quality assessment.

Stem quality attributes cannot be predicted without prior knowledge of the stand under consideration. Thus, it is assumed that some form of assessment will have to be made in the field. Usually, this has to be done at minimum cost, and a number of methods are available to carry out a stem quality assessment. The assessment techniques may be subdivided in 3 groups (Fig. 4-15).

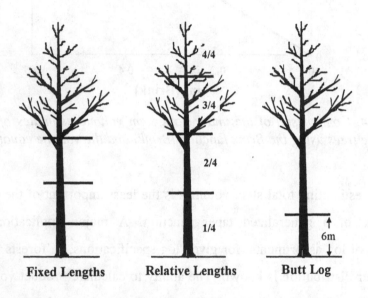

Fixed Lengths   **Relative Lengths**   **Butt Log**

Figure 4-15. Schematic representation of three methods of stem quality assessment.

The *fixed lengths* method requires subdivision of the utilizable stem section into one or more lengths, each of which is assigned to a quality type. The corresponding tree heights have to be measured or estimated, which may be costly. Based on the data and using predefined quality classes, assortment distributions are then calculated. Mikulka (1955) used three 5-m sections in mature beech stands, while the Swiss National Forest Inventory (Schweizerisches Landesforstinventar, 1988) assigns one of three possible quality classes to the first two 4-m sections of standing trees with a minimum breast height diameter of 20cm. Düser (1978) estimated utilizable logs in beech stands by sampling the lengths up to the base of the crown.

The *relative lengths* method subdivides the whole tree into sections of equal length and assigns each section to one of a fixed number of predefined quality classes. It is fairly easy to establish relative tree lengths in the field, this being the main advantage of the technique which was used by Speidel (1955), Bernstorff and Kurth (1988) and Bachmann (1990).

The *buttlog method* is based on the premise that the lower part of the stem contains a major share of the value of a tree, and that it is sufficient to limit the assessment effort to the lower 4- or 6-m section. One of the first applications of the buttlog method was the large-scale inventory carried out in valuable beech and oak stands in Northern Germany by Arnswaldt (1950).

Although being one of the most effective techniques of stem quality assessment, the buttlog method may be improved in two ways. Firstly, the volume calculations could be made more precise by using appropriate taper functions (Trincado and Gadow, 1996). Secondly, the assignment of quality classes could be based on quality criteria measured in the field, using a key to consider buyer-specific requirements (Wiegard et al., 1997).

The essence of the buttlog method is an appraisal of species-specific visible log quality criteria which are assessed in the first 4-m (or 6-m) log. For oak trees, the following five quality criteria are often used: *branches, lumps, wounds, sinuosity* and *twist*. Such visible criteria may reveal between 70 and 80% of the inner defects (Schweizerische Landesforstinventar, 1988).

Figure 4-16 presents a simple classification of the possible combinations, involving the five criteria and their possible expressions, which may be encountered in the buttlog of an oak tree. The first defect *branches* is characterized by the branch diameters and by the number of branches occurring in the buttlog. *Lumps* and open *wounds* are differentiated on the basis of their size, using threshold diameters of 8 and 10cm for defining the categories. The criterion *sweep* measures the horizontal distance between the tree and a perpendicular rod, while *twist* refers to the deviation of stem fibers, turning around the stem axis.

The column *number* in Fig. 4-16 gives frequency thresholds for the 6-m buttlog, while the column *key* presents a discrete number of categories for each of the quality criteria. The keys are used to assign quality classes to the buttlog in accordance with specific buyer requirements. The threshold values in Fig. 4-16 were derived on the basis of research done by Schulz (1954). Of course, these values need to be adapted, if necessary, for different markets, tree species and other circumstances.

One of the problems associated with the buttlog method is the undifferentiated lumping of information for the entire length of the buttlog. The height at which a defect occurs is not assessed. It may thus happen, for example, that a 6m buttlog of good quality contains only one serious defect at a height of 5m which would lead to a declassification of the otherwise excellent 5-m section. The lumping problem may cause underestimates of product yields of optimum quality, which are degraded as a result of a defect at the extremes of the buttlog, and overestimates of yields of poor-quality logs. Average quality logs may be under- or overestimated.

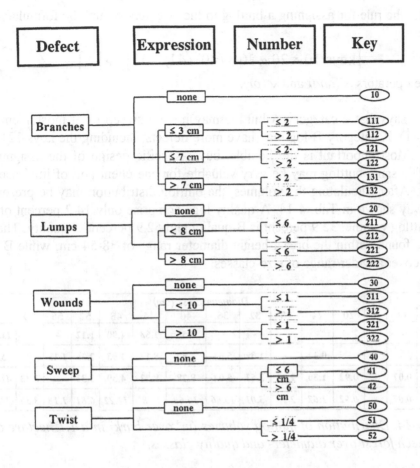

Figure 4-16. Quality criteria used by Wiegard (1996) in applications of the 6-m buttlog method in valuable oak stands.

The buttlog method may, of course, be adjusted if a more accurate quality assessment is required, for example, in very valuable stands simply by further subdividing the buttlog. An important extension of the method would be to establish a statistical relationship between the buttlog defects and those found higher up.

Example: The buttlog method is used to obtain the distribution of the buttlog volumes (in m^3 over bark) of the 124 valuable oak trees in a 118 -year old oak-beech forest covering an area of 1.1 hectare. After completion of the field work, the buttlogs are assigned to quality classes using a specific key for each class. For example, the rule for assigning a buttlog to the A-category may be formulated as follows:

$$\{keys: 10 \wedge 20 \wedge 30 \wedge (40 \vee 41) \wedge 50\} \Rightarrow [A]$$

with the operators $\wedge$ *(and)* and $\vee$ *(or)*.

The rule says that a category A-buttlog may have a sweep of at most 1 cm per meter (41). A category B-log may have more defects, including the keys 121, 42 and 51. Most important is the flexible, buyer-specific design of the assignment rules. The same buttlog may be very valuable for one client and of little use for another. After calculating the volumes, the buttlog distribution may be presented in the way shown in Tab. 4-14. A-quality logs comprise only 14.2 percent of the total buttlog volume, 32.9 percent is B-quality and 52.9 percent C-quality. The A-logs are found within the breast-height diameter range of 38-54 cm, while B and C-logs cover a wider range of size-classes.

Quality	Diameter class (cm)													
Class	12	16	20	24	28	32	36	40	44	48	52	56	60	Total
A	-	-	-	-	-	-	-	3.58	1.53	4.90	1.12	-	-	11.12
B	-	-	-	0.32	-	1.38	5.64	4.82	8.12	1.93	2.36	1.18	-	25.73
C	0.07	-	0.92	1.30	2.18	3.63	5.02	8.28	9.22	4.39	3.33	-	3.33	41.34
Total	0.07	-	0.92	1.62	2.18	5.01	10.66	16.68	18.87	11.22	6.81	1.18	3.33	78.19

Table 4-14. Distribution of buttlog volumes (m³ over bark) in a 1.1 hectare oak-beech forest over diameter- and quality classes.

Quality assessments are a key issue in forests, where the timber price is mainly influenced by value, for example, in many oak and beech forests, but also in some valuable pine stands. Harvesting decisions can be made on the basis of value increment, and foresters are able to respond to the fluctuating timber market. An inventory concept known as *silviculture controlling* may be used to assess the changes in the size and value distributions, which are caused by a thinning (Gadow, 1995).

Modelling Thinnings

In a managed forest, thinnings usually have a far greater effect on forest development than natural growth. Although the prediction of the diameter distribution after thinning is an important problem, surprisingly little has been published on this particular topic, when compared to the wealth of literature covering tree growth. Over the years, the problem has been adressed using tables or regressions involving various stand parameters and variables describing the weight and type of thinning. For example, a table compiled by Craib (1939) was used by several authors to calculate thinning effects in South African plantations. A case in point is Grut (1970), who derived regression equations for predicting the change of the mean diameter after a thinning in *Pinus radiata* plantations. Similar methods were used by Kennel (1972), Preussner (1974), Alder (1979), Römisch (1983), Gadow (1987), Murray and Gadow (1991), Lemm (1991) and Kassier (1993). The most common approach is to describe a thinning in terms of the change of the diameter distribution.

Change of Distribution Parameters

One of the methods for assessing the change is based on the estimation of the effect of a thinning on the parameters of the diameter distribution.

An example of this approach is provided by Álvarez González (1997) who estimates the change of the Weibull parameters b und c after a thinning in *Pinus pinaster* forests in Spain:

$$b_{after} = -4.7067 + 1.0205 \cdot b_{before} + 85.35 \frac{N_{removed}}{N_{total}} - 73.617 \frac{G_{removed}}{G_{total}} \qquad 4\text{-}39$$

$$c_{after} = -1.059 + 1.178 \cdot c_{before} + 8.170 \frac{N_{removed}}{N_{total}} - 5.255 \frac{G_{removed}}{G_{total}} \qquad 4\text{-}40$$

with b_{before}, b_{after}, c_{before} and c_{after}=Weibull parameter b and c before and after a thinning,

$N_{removed}$, N_{total}, = number of stems per ha, removed and before thinning,
$G_{removed}$, G_{total}, = basal area per ha, removed and before thinning.

The removed stem number and basal area ratios are very suitable independent variables for modelling thinning effects (cf. Chapter 4).

An alternative approach was developed by Gadow and Hui (1997) using data from the long-term *Cunninghamia lanceolata* thinning trials *DGS* and *CX*, established in China. *DGS* was initiated in 1988, and low thinnings were carried out immediately after its establishment. Between 15 and 65 percent of the stems per ha were removed. Fig. 4-17 shows an example of the diameter distribution before and after the thinning in Plot 310 of the trial *DGS*.

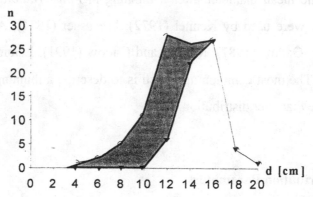

Figure 4-17. Example of a diameter distribution before and after thinning at age 12, observed in Plot 310 of the growth trial DGS. The shaded area indicates the removed trees.

The thinning trial *CX*, comprising 12 plots, was initiated in 1975, at age 11 years, by the Forestry Institute in Jiangxi. One low thinning was carried out in 1975 at age 11. The plots were enumerated again 1, 2, 5, 8, 10 and 14 years after the thinning.

An example of the movement of the cumulative diameter distribution between 1975 and 1986 is presented in Fig. 4-18, using the data obtained in Plot 7 of the trial *CX*.

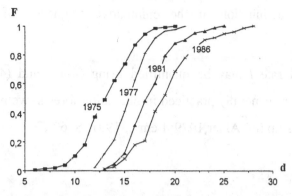

Figure 4-18. Example of the movement of the cumulative diameter distribution from 1975 to 1986, observed in Plot 7 of the thinning trial CX.

The survival rate of a representative tree in a diameter class is, according to Alder (1979), determined by its relative cumulative frequency. The remaining trees per ha after a thinning may be obtained according to:

$$nr_i = N \int_{F(x_{i-1})}^{F(x_i)} l \, dF$$

4-41

with nr_i = stems per ha remaining after thinning in diameter class i.
                N = stems per ha before thinning,
     $F=F(x)$ = relative cumulative diameter frequency before thinning,
               x = diameter [cm],
                l = survival rate of a tree with diameter x.

The total survival rate L, i.e. the relative proportion of the trees remaining after the thinning is:

$$L = \int_{0}^{F=1} l(F(x)) \, dF$$

4-42

The thinning weight in terms of stems per ha is thus *(1-L)*. Equation 4-42 can be used to calculate the diameter distribution after a thinning, if the distribution before the thinning and the survival rate l as a function of the relative cumulative frequency are known. Thus, one way to model a thinning is to estimate the survival rate as a function of the cumulative frequency of the diameter distribution.

The survival rate l may be quantified using (4-42) and (4-43). For low thinnings, which are generally practiced in plantation forests, we may specify the following relationship (cf. Alder,1979; Lemm, 1991, S. 62 ff):

$$l(F) = F^c \qquad\qquad\qquad 4\text{-}43$$

Using (4-42) we obtain:

$$L = \int_0^1 F^c dF = \frac{1}{c+1} F^{c+1} \Big|_0^1 \qquad\qquad 4\text{-}44$$

Thus:

$$L = \frac{1}{c+1} \ \text{with}\ c = \frac{1}{L} - 1 \ \text{and}\ l(F) = F^c \qquad 4\text{-}45$$

The relationship between F and l is shown in Fig. 4-19 for different thinning weights $1\text{-}L$.

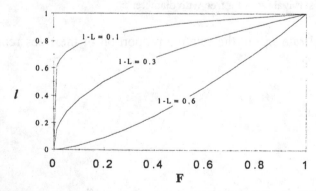

Figure 4-19. Relationship between the cumulative diameter frequency F and the survival rate l for varying thinning weights (1-L).

Thus, the probability that a tree survives a thinning increases with increasing diameter. The degree of selectivity is influenced by the weight of the thinning.

The removal rate is very high in the smaller diameter classes, when the thinning is light, but it is more evenly distributed over the diameter classes when the thinning is heavy.

From (4-41) we obtain:

$$nr_i = N \int_{F(x_{i-1})}^{F(x_i)} F^c dF = N \frac{1}{c+1} F^{c+1} \Big|_{F(x_{i-1})}^{F(x_i)} \qquad 4\text{-}46$$

The diameter distribution after a thinning, for a given thinning weight, is equal to:

$$nr_i = NL\left[F(x_i)^{\frac{1}{L}} - F(x_{i-1})^{\frac{1}{L}} \right] \qquad 4\text{-}47$$

$F(x_i)$ refers to the stand before thinning. When a thinning does not occur, we have $L=1$ and $nr_i=N\{F(x_i)-F(x_{i-1})\}$.

The method was applied using the data of the *DGS* thinning trial. The theoretical and observed diameter distributions were compared for one thinning in all the plots (Tab. 4-15).

No	1-L	χ^2	f	χ^2(f,0.05)
103	0.454	7.997	4	9.488
104	0.542	4.424	5	11.070
106	0.600	0.150	3	7.815
107	0.500	9.018	4	9.488
108	0.368	6.476	4	9.488
109	0.250	4.212	5	11.070
110	0.265	6.638	5	11.070
111	0.641	5.529	3	7.815
112	0.593	4.418	4	9.488
301	0.191	3.898	5	11.070
302	0.492	8.126	4	9.488
303	0.515	2.720	4	9.488
305	0.435	4.667	6	12.592
306	0.446	5.492	6	12.592
310	0.428	8.166	5	11.070

Table 4-15. Results of the χ^2-goodness-of-fit test for comparing the theoretical and observed diameter distributions of the remaining stand in the thinning trial DGS (Nr=plot number; 1-L=thinning weight (relative stem number removed); χ^2: calculated chi-square value; f: degrees of freedom; χ^2(f.0.05): χ^2-table value at the 5%-level).

In all the plots, the calculated χ^2 values are less than the table values at the 0.05 level. The model thus performs well in estimating the thinning effects on the distributions of the *Cunninghamia lanceolata* trials.

The simplicity of the method, which appears to work well in regular, even-aged plantation forests, is appealing.

Movement of the Diameter Distribution after Thinning

The development of a diameter distribution after a thinning is normally modelled using the parameters of a distribution function, which are estimated as a function of age and stand attributes or various methods of stand table projection (see previous sections). A very simple approach was used by Hui and Gadow (1996), using the following equation for describing the cumulative distribution function:

$$F = \frac{1}{1 + \exp(a - bX)} \qquad\qquad 4\text{-}48$$

For a *Cunninghamia lanceolata* plantation, the parameters may be estimated directly from the percentile values ($X_{F=0.5}$ and $X_{F=0.9}$) using the following equations:

$$a = bX_{F=0.5} \quad \text{with} \quad b = \frac{2.19722}{X_{F=0.9} - X_{F=0.5}}$$

The cumulative diameter distribution is defined by the percentile values Dx (where $Dx = X_{F=0.5}$ or $Dx = X_{F=0.9}$) and the change of the two percentile values describes the movement of the diameter distribution. We assume that the ratio of the differences between the logarithms of a percentile value and the quadratic mean diameter at two points in time is constant:

$$\ln Dx_2 - \ln Dg_2 = k(\ln Dx_1 - \ln Dg_1)$$

Therefore:

$$Dx_2 = Dg_2 \left(\frac{Dx_1}{Dg_1} \right)^k \qquad\qquad 4\text{-}49$$

The following parameter estimates were obtained using the data from the thinning

experiment CX:

Dx	k	R²	MSE
$X_{F=0.5}$	1.11904	0.978	0.072
$X_{F=0.9}$	1.02653	0.980	0.112

The model was validated using the Kolmogoroff-Smirnoff test and the data of the

thinning experiment DGS, which was remeasured 6 years after the last thinning.

The results are summarized in Tab. 4-16.

No	Nn	Pn	Dgn	Dg6	K-S	D(n,0.05)
103	1380	0.454	14.3	17.2	0.132	0.185
104	1380	0.542	14.5	17.2	0.079	0.175
106	1050	0.600	14.3	16.6	0.037	0.196
107	1335	0.500	14.9	17.8	0.052	0.177
108	1560	0.368	14.7	16.5	0.015	0.161
109	1545	0.250	13.8	16.0	0.076	0.172
110	1530	0.265	14.4	16.6	0.064	0.160
111	1050	0.641	14.0	17.4	0.097	0.198
112	1095	0.593	13.8	16.5	0.084	0.196
301	1560	0.191	13.6	16.3	0.084	0.160
302	1320	0.492	14.7	17.8	0.020	0.180
303	1050	0.515	15.8	18.8	0.106	0.214
305	1050	0.435	16.0	18.2	0.117	0.196
306	1560	0.446	14.6	17.5	0.043	0.160
310	1380	0.428	15.1	17.6	0.056	0.175

Table 4-16. Results of the validation using the K-S-test. K-S: maximum absolute difference between observed and theoretical distribution; D(n,0.05): critical K-S-value at α=0.05 level of significance. No: trial number; Nn: stems per ha after thinning; Pn: relative stem number removed; Dgn: mean squared diameter immediately after thinning (cm); Dg6: mean squared diameter 6 years after thinning (cm).

The method is simple and very effective for predicting the movement of the

diameter distribution in an even-aged timber plantation.

The following numerical example is added for further clarification:

Example: Hypothetical initial diameter distribution at the age of 12 years:

D-class midpoint x_i (cm)	4	6	8	10	12	14	16	18	20
number of trees n_i	1	2	5	11	28	26	27	4	1

A low thinning removes 42.8% of the trees, thus L=0.572. Using Eq. (4-47) we obtain the diameter distribution after thinning:

D-class Midpoint x_i (cm)	4	6	8	10	12	14	16	18	20
Number of Trees Remaining nr_i			1	2	12	17	23	4	1

The following diameter distribution is obtained at the age of 18 years, 6 years after the thinning (Dg=17.6cm):

D-class Midpoint x_i (cm)	4	6	8	10	12	14	16	18	20	22	24
Number of Trees Remaining nr_i					1	6	18	22	8	10	1

Separation Parameters

Certain stand characteristics, specifically the mean and the variance of the diameter distribution, are presumed known before a thinning. The objective is then to describe the diameter distributions of the removed and the remaining trees when the thinning proportion, defined as the proportion of trees which are removed, is given. Intuitively, the degree of selectivity, i.e. the preference for removing trees in certain diameter classes rather than at random, should be relevant when describing the relationship between the distributions. Several authors have found specific regression equations for predicting the changes of the distribution parameters following a thinning, but neither the values of the coefficients nor the particular form of the equation is necessarily applicable in other circumstances. In fact, there is no clear methodology of how such equations should be derived for other data sets.

In non-selective thinnings, such as when every n'th row is removed, the mean diameters before and after the thinning are expected to be the same. For modelling purposes, one can therefore make the assumption that the mean diameter does not change during a non-selective thinning. Clearly, no additional distributional information is required. In a selective thinning, however, the mean diameter does change. The spread in tree sizes permits selection of trees with small (or large) diameters while the larger (or smaller) trees remain. Consequently, some additional information about the distribution is essential. At least the variance must be known. Thus, Murray and Gadow (1991) focussed their attention on the relationships between the means and variances of the before-thinning, the removed and the remaining trees and the consistency that must be maintained between the three diameter distributions.

Let μ and σ^2 be the mean and variance of the diameter distribution, while the subscripts b, t and r distinguish between the before-thinning distribution, the removed distribution and the remaining distribution. The proportions of the number of trees removed and remaining are denoted by a_t and a_r, respectively. Obviously the condition:

$$a_t + a_r = 1 \qquad\qquad\qquad\qquad 4\text{-}50$$

must hold and is equivalent to

$$N_t + N_r = N_b \qquad\qquad\qquad\qquad 4\text{-}51$$

with N_t, N_r, and N_b denoting the numbers of trees.

In terms of this notation, the before-thinning quantities μ_b and σ^2_b are assumed to be known, while the thinning proportion a_t is given. The problem is to determine μ_r and μ_t, and σ^2_r and σ^2_t from these known quantities.

A thinning splits the diameter frequency distribution into two parts, the removed and the remaining part, and it is essential that the two parts add up to the total. The associated consistency conditions imply very specific forms for the equations, relating the quantities μ_r and μ_t, and σ^2_r and σ^2_t to their before-thinning counterparts μ_b and σ^2_b and the thinning proportion a_t. The consistency conditions will focus on the general structure of the model.

In addition, an empirical element is required for describing the type of thinning encountered in practice. The selectivity of a thinning operation can be described by the difference between the means of the removed and the remaining distributions. This quantity can be made dimensionless by dividing it by a scale parameter, e.g., the standard deviation of the original distribution. The *location separation parameter* S_1 is thus defined as:

$$S_1 = \frac{\mu_r - \mu_t}{\sigma_b} \qquad\qquad\qquad 4\text{-}52$$

Equation (4-52) may be rewritten as

$$\mu_r - \mu_t = S_1 \cdot \sigma_b \qquad\qquad\qquad 4\text{-}53$$

The mean of the original distribution is the weighted average of the mean of the thinned and the remaining distribution, which is embodied in the condition[1]

$$a_r \mu_r + a_t \mu_t = \mu_b \qquad\qquad\qquad 4\text{-}54$$

Solving for μ_r and μ_t in (4-53) and (4-54) yields

$$\mu_r = \mu_b + a_t \sigma_b S_1 \qquad\qquad\qquad 4\text{-}55a$$

and

$$\mu_t = \mu_b - a_r \sigma_b S_1 \qquad\qquad\qquad 4\text{-}55b$$

Thus, S_1 is the only empirical quantity required to determine the thinned and the remaining mean diameters in terms of the before-thinning parameters μ_b and σ^2_b and the thinning proportion a_t (actually a_r is also used, but $a_r = 1 - a_t$).

Analogous to the preceding analysis, a consistency condition may be derived relating the variances of the three distributions. First, we define the *scale separation parameter* S_2:

$$S_2 = \frac{\sigma^2_r - \sigma^2_t}{\sigma^2_b} \qquad\qquad\qquad 4\text{-}56$$

---

[1] An equivalent equation was used by Nagumo et al. (1988) to compute μ_r from the other variables, but this is only feasible when μ_t is already known.

The following consistency conditions can be derived, based on classical sums of squares conditions in the analysis of variance[2]:

$$\sigma^2_r = \sigma^2_b(1 - a_r a_t S_1^{\,2} + a_t S_2) \qquad\qquad 4\text{-}57a$$

and

$$\sigma^2_t = \sigma^2_b(1 - a_r a_t S_1^{\,2} + a_r S_2) \qquad\qquad 4\text{-}57b$$

These equations express the variances of the remaining and of the removed distributions in terms of the before-thinning variance, the thinning proportions and the separation parameters.

Note that no assumptions have been made regarding the form of the distributions. For any particular distribution, the first two moments of the thinned and the remaining distributions are available from Equations (4-55) and (4-57). Third moments may be found by defining a skewness separation parameter and imposing specific consistency conditions. The Weibull distribution parameters can then be obtained using methods based on moments (Burk and Newberry, 1984; Shiver, 1988).

Example: Fig. 4-20 (which is presented again for convenience) shows the results of two thinning exercises in a beech forest. A low thinning and a high thinning were simulated in a sample plot covering an area of 0.1. ha.

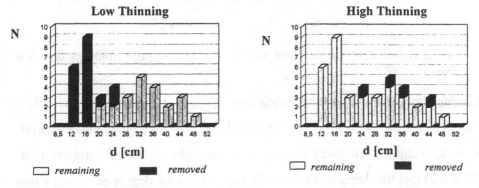

Figure 4-20. Results of a low thinning and a high thinning in a 0.1 ha plot in a beech forest in Northern Germany.

[2] For details of the derivation, refer to Murray and Gadow (1991).

The thinning weight in terms of the basal area removed was approximately equal. The values of the location separation parameter are $S_1=1.7$ for the low thinning and $S_1=-0.74$ for the high thinning.

The practice of silviculture is characterized by numerous semantic expressions, which are used to describe a particular thinning operation. These descriptions are not very precise and thus a particular expression, such as *high thinning,* may be associated with a range of different S_1-values. It is possible though, to assign an S_1-parameter value to a semantic expression with different degrees of association. To this effect, we may use an *association function,* which defines the strength of association of a given S_1-value to one or more categories (Zimmermann, 1991). Fig. 4-21 shows an association function for the two categories *high thinning* and *low thinning.* The functions are based on an analysis of numerous long-term thinning experiments in beech forests in Northern Germany and Denmark.

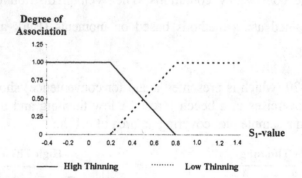

Figure 4-21. S_1-*association function for the categories high thinning and low thinning.*

A thinning may be classified as definitely *low,* given S_1-values greater than 0.8; if the S_1-value is less than 0.2 we may be sure that it was a high thinning. However, with an S_1-value of 0.6, there is only a one-in three-degree of certainty (0.8-0.6)/0.8-0.2) that the thinning was actually *high,* while the degree of certainty that it was a *low thinning* is equal to (0.6-0.2)/0.8-0.2)=2/3. The example again shows the problems associated with the use of semantic variables for describing silvicultural operations, and the opportunities for improving the descriptions, not only in the aftermath of a thinning, but also in the planning stage.

Modelling Foresters' Tree-Selection Behavior

Selecting trees for removal is one of the most important tasks in the small-scale *forest gardening* systems practiced in Central Europe. This task is increasingly delegated to a harvesting machine operator, assuming that the resulting modifications of the stand structure are similar or even identical to those effected by a professional forester. Thus, the question arises: are there differences in the "*marking behavior*" between different foresters, or between a forester and a harvesting machine operator, who are given the same set of silvicultural objectives? If there are differences, then it would not be sufficient to model average thinning effects. One would need to model the individual tree *marking behavior* of each forester or machine operator who is engaged in selecting trees for removal.

The first problem that needs to be addressed is the method of measuring differences in tree-selection patterns. As an example, consider an exercise involving four teams of experienced foresters, who were asked to mark a forest stand comprising 300 trees, principally *Abies alba* and *Fagus*. Each team was given the same set of objectives outlining the silvicultural principles and specific aims of the thinning. They then proceeded to mark the stand by indicating the suggested removals on a numbered tree list. The results of the trial thinnings are available as a list of 300 trees, together with the number of foresters who marked each of the trees, that is the number of marks, out of four, assigned to each tree. The results are summarised in Tab. 4-17.

Number of Marks	0	1	2	3	4
Frequency	191	70	30	5	1

Table 4-17. Frequency distribution of the number of marks on 300 trees.

Thus, for example, 191 of the 300 trees were not marked by any of the four foresters, only one was marked by all four foresters. The problem, finding a way of quantitatively assessing the degree of agreement or disagreement between the foresters, was addressed by Zucchini and Gadow (1995). As the responses of the individual foresters are not available, it is not possible to make comparisons between the marking behavior of individual foresters. For example, it is not possible to determine whether differences that exist might be largely attributed to one of the four foresters. Thus, only indices of overall agreement can be used.

In general, an index is a measure designed to facilitate comparison between a given situation and one or more standard situations that provide interpretable reference points. For example, the correlation coefficient which measures the strength of the linear relationship between two variables has three reference points, namely 1, 0 and -1, which indicate, respectively, a perfect linear relationship with positive slope; no linear relationship, and a perfect linear relationship with negative slope.

For the purpose of interpretation, the most important consideration in the construction of an index of agreement has to do with the choice of reference points, that is how the index is standardized. It is possible to regard the frequencies in Tab. 4-16 from very different points of view. In order to reflect a particular point of view using an index, it is necessary to standardize the latter accordingly. It would seem that the general type of index that is employed does not materially effect the assessment of overall agreement, but the way in which the index is standardized does.

Suppose there are K trees and that each of n foresters either mark or do not mark each tree. Let x denote the number of foresters that mark a particular tree, and $(n-x)$ those that do not mark it. The absolute difference between x and $n-x$ provides a measure of agreement for a given tree. For $n=4$ foresters, the possible values of the absolute difference $|x - (n - x)| = |2x - n|$ are listed in Tab. 4-18.

x	n - x	$|2x - n|$
0	4	4
1	3	2
2	2	0
3	1	2
4	0	4

Table 4-18. All possible values of absolute differences, given four foresters.

Suppose that the trees are numbered $k = 1,2,...,K$ and that x_k denotes the number of foresters that marked tree k. Then:

$$C = \sum_{k=1}^{K} |2x_k - n|$$

can be used to construct an index of agreement among the foresters for the stand of trees under consideration. The minimum and maximum possible values of C are, respectively, 0 and $C_{max} = nK$. A possible index is thus:

$$I_1 = \frac{C}{C_{max}}$$

There is no difficulty in interpreting the upper bound of this index; if $I_1 = 1$ then there is complete agreement. However, the interpretation of the lower bound (0) is less straightforward. Note that in the case of $n=2$ foresters, the lower bound is achieved if there is *perfect disagreement*, i.e. the two foresters assess every single tree in the opposite way. In other words, $I_1 = 0$ implies a greater degree of disagreement than would occur if, for example, the foresters were to select the trees to be marked *at random*.

We will, therefore, make a distinction between the terms "*disagreement*" and "*lack of agreement*". The former will be used to describe a situation in which the foresters are reaching *contrary conclusions* and the latter for a situation in which their *conclusions are unrelated*.

In view of this interpretation, it may be preferable to standardize the index in such a way that a situation in which the selection is made independently and at random receives the value 0. One such index is:

$$I_1(p) = \frac{C - C_{ind}(p)}{C_{max} - C_{ind}(p)}$$ 4-58

where $C_{ind}(p)$ is the expected value of C under the condition that the foresters select the trees to be marked independently and with probability p. The expectation C_{ind} is computed using the binomial probability function:

$$C_{ind}(p) = K\sum_{x=0}^{n} |2x - n| \binom{n}{x} p^x (1-p)^{n-x}$$ 4-59

No *assumptions* are made about the behavior of the foresters. Only a different baseline is set for the index. If each forester *were* to mark trees at random with probability p, then the expected value of $I_1(p)$ would be zero. Values of $I_1(p)$ that are less than zero reflect *disagreement rather than lack of agreement.*

The value of p has a substantial influence on the index, because it determines its zero level and hence its interpretation. Ideally, p should be set equal to the proportion of the trees in the stand that the foresters were supposed to mark, if the proportion was (either explicitly or implicitly) pre-assigned. For example, it might have been understood by the foresters that approximately 15% of the trees should be marked, in which case the appropriate index would be $I_1(0.15)$.

Tab. 4-19 gives the observed frequencies with the corresponding expected frequencies obtained in the scenario that each forester uses a marking strategy, in which each tree is marked independently and at random with probability $p = 0.15$.

No. of Marks	0	1	2	3	4
Observed Freq	191	70	30	5	1
Expected Freq	156.6	110.5	29.3	3.4	0.2

Table 4-19. Observed and expected frequencies for the number of marks (p = 0.15).

Note that 191 out of 300 trees were marked by none of the foresters. This might suggest a high degree of agreement. However, the corresponding expected frequency under the hypothetical scenario in which the foresters mark about 15% of the trees at random, is 156.6, indicating that the degree of agreement is not as high as it might appear to be at first glance. Thus, in deciding how to assess the degree of agreement, it is important to specify the reference point to be used. For the frequencies in Tab. 4-17 one obtains:

$$I_1 = 0.77$$
$$I_1(0.15) = 0.20.$$

The gross discrepancy between the values of the two indices reflects the very different assumptions on which they are based. The second index is appropriate in situations where the proportion of trees to be marked is (perhaps implicitly) prescribed *a priori* to be approximately 0.15. From this perspective, there is almost a lack of agreement but, since $I_1(0.15)$ is greater than zero, there is no indication of disagreement.

The first index, I_1, would be appropriate if each forester felt entirely free to mark as many or as few trees as they wished, that is with absolutely no prior notion of the percentage to be marked. From that perspective, as indicated by the high value (0.77), the degree of agreement is high.

In cases where p is not known, but in which it is known that such a value was either actually pre-assigned or implied, it would be reasonable to use the index $I_1(\hat{p})$ where

$$\hat{p} = \left(x_1 + x_2 + ... + x_k\right) / nK,$$

which is the average proportion of trees marked by the n foresters. Here $\hat{p} = (191 \times 0 + 73 \times 1 + 30 \times 2 + 5 \times 3 + 1 \times 4) / (4 \times 300) = 0.127$. The associated index value is $I_1(0.127) = 0.06$ which, again, indicates that there is lack of agreement, but no disagreement.

It is necessary to keep in mind that the assessment refers to the performance of the foresters in respect to a particular stand. Some stands will be such that the trees are easier to classify and thus such that the same four foresters would be in complete agreement. Conversely, if the trees are difficult to classify, then there will tend to be fewer agreements. Thus, the variation of tree attributes, that is the degree of homogeneity in the stands, needs to be considered.

Now consider a similar exercise involving 5 foresters. This time, the attributes of the 300 numbered trees are available. Thus, for evaluating the results, it is possible to use a generalized linear model for binary variables (Füldner et al., 1996). Two options *remove* or *remain* are available for each tree. The decision is influenced by a number of factors (also known as *covariates*) such as the intrinsic attributes of the tree itself and the neighbourhood constellation, including the attributes of the competitors in the immediate vicinity of the tree. This influence must be quantified.

First we define Z_1, Z_2, ...Z_n to be the destinies of the trees in the stand with:

$$Z_i = \begin{cases} 1, \text{if tree i is removed} \\ 0, \text{if tree i remains, with i} = 1, 2, ..., n. \end{cases}$$

In addition, the variable $\pi_i = P\{Z_i = 1\}$ is introduced, which represents the probability that tree i is removed.

There are two decision alternatives *remove* or *remain*, and the outcome is comparable to throwing a coin. Heads means remove, tails remain. However, the coins are not identical. A different coin is thrown for each tree, because the probability that *heads* is thrown is a particular one for the i'th tree, namely π_i. If, on account of its attributes, it is undesirable that tree i remains, the value of π_i will be closer to 1 with a higher probability that the coin shows *heads*. However, if the tree, on account of its attributes, should remain, then the coin will be biased towards *tails*.

The random variables (Z_i) can only assume one of two values (0 or 1). Thus, the commonly used linear models, such as regression analysis, are not applicable for specifying a relationship between π_i and a tree attribute. The suitable approach in this case is a so-called *generalized linear model* (McCullagh and Nelder, 1985). First, a *LINK*-function is defined for the π. The *logit link* function, for example, may be written as follows:

$$\pi_i = \frac{e^{\lambda_i}}{1 + e^{\lambda_i}}, \quad i = 1, 2, \ldots, n \qquad\qquad 4\text{-}60$$

with $\lambda_i = \beta_o + \beta_1 D_i$, $i = 1, 2, \ldots n$, defining a linear relationship between the tree diameter D_i and the transformed values of π_i. The parameters β_0 and β_1 can be estimated using available statistical software (such as *S-Plus*, *GLIM* or *SAS*; see McCullagh and Nelder, 1985 and Dobson, 1990, for details about theory and algorithm). The parameter values are listed in Tab. 4-20.

Forester	β_0	β_1
A	-3.213	0.03543
B	-2.178	0.00312
C	-2.023	0.02330
D	-2.313	0.00739
E	-2.013	0.01639

Table 4-20. Parameter values for the logit link *function.*

A relationship between the tree diameters and the selection probabilities is shown in Fig. 4-22 for each of the five foresters. There are obvious differences among the foresters relating to their selection probabilities. The foresters D and B do not show any diameter preference. Their selection probability is indifferent to the tree diameter, amounting to about 10% in any tree size. The selection probability of foresters A, C and E increases with increasing tree diameter. Forester A has a very strong preference for the big trees. Furthermore, it is noticeable that the foresters apply different thinning weights. C takes out considerably more trees than B, for example.

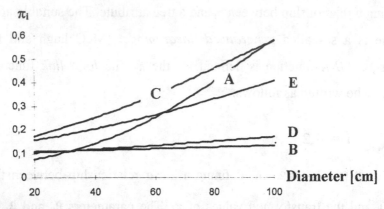

Figure 4-22. *Relationships between the tree diameters and the selection probabilities for the five foresters.*

The example shows that considerable person-specific variability of thinnings may occur, given the same silvicultural objectives[3]. This observation leads to several conclusions and considerations. Firstly, the commonly used silvicultural specifications for describing the objectives of a thinning are obviously not operational (otherwise they would all have had the same selection probabilities). Secondly, it may be desirable from an ecological perspective to allow such variations in the tree selection probabilities, reflecting the differences in the individual experiences of the foresters and leading to greater diversity of forest structures. Thirdly, if such variability is permitted or even encouraged, then it should be made transparent. Otherwise, it will be impossible to forecast forest development, which is not only influenced by natural growth, but also by the thinnings. Thus, models describing individual forester-specific modifications of forest structures are an essential part of any growth modelling effort.

---

[3] However, Kahle (1995) found a great degree of consistency displayed by the same forester who was asked to simulate a thinning in the same stand several times in monthly intervals.

Chapter Five

Individual Tree Growth

Regional yield models, used for large-scale resource forecasting, estimate the standing volume of a timber resource for different ages, site qualities and degrees of stocking. *Stand models* predict average tree dimensions and area-based quantities for whole stands, while *size-class models* project the development of a limited number of representative trees, which possess average attributes within a cohort of similar trees.

Distance-dependent individual tree-growth models maintain the identity of each tree, resulting in an increased level of modelling resolution. Furthermore, it is assumed that coordinates are available and thus the immediate neighborhood constellation is known for each tree (Fig. 5-1). This additional information, which is usually obtained in remeasured sample plots (Akça, 1993), considerably widens the scope for a variety of techniques that depend on known tree positions and tree-to-tree distances. Measuring tree coordinates may be laborious, and this information is rarely available in practice. However, a precondition to using a position-dependent individual tree model is the ability to generate the tree positions. This is a task of central significance. Once the positions and their attributes are known, competition indices may be calculated and position-dependent growth models applied. Thinnings may also be described and modelled in much greater detail.

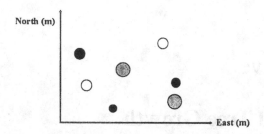

Figure 5-1. Hypothetical section of a forest with 7 trees with known coordinates and breast height diameters. Shading indicates different species.

The tree coordinates and their attributes are usually obtained from permanent sample plots, but may be generated automatically from digitized aerial photos (Dralle, 1997). However, this technique has limitations and in many circumstances, the method does not yet produce satisfactory results. Alternatively, the tree coordinates and attributes may be obtained using simulation techniques as applied, for example, by Pretzsch (1994) and Lewandowski and Gadow (1997).

Generating Spatial Structures

Tree coordinates and their attributes may be generated using specific simulation techniques. The objective of *reconstructing* a forest in this way is to (a) improve monitoring of silvicultural operations and (b) to use more advanced growth models based on known neighborhood constellations. A reconstruction is considered successful if there is a close resemblance between the real forest and the artificial one.

Specific spatial information is needed to measure the resemblance between the real forest and the reconstructed one. The necessary spatial information is provided by variables describing the relationships among neighboring trees. Such variables represent a useful basis for reconstructing the spatial structure of a forest.

Figure 5-2. Uneven-aged forest with beech (Fagus sylvatica), Spruce (Picea abies) and Douglas Fir (Pseudotsuga menziesii) in Northern Germany.

Variables for Describing Spatial Structure

A convenient way for describing the spatial structure of a forest is to use the attributes of neighboring trees on the levels of size, species and distance. While size *differentiation* quantifies the differences in the sizes of neighboring trees, *mingling* refers to the way in which neighboring trees of different species intermingle, and *aggregation* measures the distribution of tree-to-tree distances.

Size Differentiation

A characteristic attribute of a forest is the differentiation of the sizes of neighboring trees. To measure this quantity, we use the variable *differentiation* introduced by Gadow and Füldner (1995). The differentiation of breast height diameters (T_i) for a given tree i (i=1...I) and its n nearest neighbors j (j=1..n) is defined as follows:

$$T_i = 1 - \frac{1}{n}\sum_{j=1}^{n}\frac{\min(d_i, d_j)}{\max(d_i, d_j)}$$

5-1

with d = breast height diameter (cm)[1] and $0 \le T_i \le 1$.

The principle is illustrated in Fig. 5-3. The breast height diameters of the i'th tree and its three nearest neighbors are given. Thus, T_i is equal to 1- 20/40 when only the first neighbor is considered, and (1- 20/40 + 1- 40/60)/2 when the two nearest neighbors are taken into consideration.

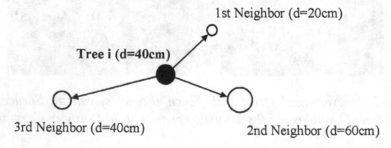

Figure 5-3. Hypothetical neighborhood constellation involving a reference tree i and its three nearest neighbors with their breast height diameters.

---

[1] It may, of course, be preferable to use other size variables, such as height, crown length or crown volume.

The value of T_i increases with increasing average size difference between neighboring trees. Zero differentiation means that neighboring trees have equal size. The index has interpretable reference points and is suitable for use by managers who need to describe forest structures using simple expressions. Of particular interest are the distributions of the T_i - values, either for generating specific structures (Lewandowski and Gadow, 1997) or random forests (Pommerening, 1997; Pommerening et al., 1997).

Distributions of size differentiation may be established for the stand as a whole or for certain sub-populations. Füldner (1995) used *species-specific* size differentiation as well as the size differentiation relating to the dominant trees.

Species Segregation and Mingling

A second spatial variable of interest describes the mingling of different tree species occuring in a forest. The index of *segregation* proposed by Pielou (1961) has been widely used. This index is based on the ratio between the observed and the expected number of mixed pairs. A *mixed pair* denotes a tree of one species (A) having a tree of the opposite species (B) as its nearest neighbor. Pielou's index of segregation for a population with two species is defined as follows:

$$S = 1 - \frac{observed \text{ number of mixed pairs}}{expected \text{ number of mixed pairs}} \qquad 5\text{-}2$$

The quantities required for calculating the index are the observed relative frequencies of one species having another as nearest neighbor, and the relative proportions of the number of stems of the two species, as shown in Tab. 5-1.

	Nearest Neighbor		
	Species A	Species B	Total
Species A	p_{AA}	p_{AB}	r_A
Species B	p_{BA}	p_{BB}	r_B
Total	s_A	s_B	1

Table 5-1. Quantities required for calculating the index of segregation for two species A and B. The p_{AA}, p_{AB}, p_{BA} and p_{BB} are the observed relative frequencies of one species having the other one as nearest neighbor; r_A, and r_B are the proportions of A and B's stem numbers; s_A, and s_B are the proportions of any tree having an A or B as its nearest neighbor.

Using the quantities specified in Tab. 5-1, the expected number of mixed pairs is equal to the sum of the joint probabilities $r_A s_B + r_B s_A$, so that the definition in Eq. 5-2 becomes:

$$S = 1 - \frac{p_{AB} + p_{BA}}{r_A s_B + r_B s_A} \qquad\qquad 5\text{-}3$$

S takes on a negative value when the observed number of mixed pairs exceeds the expected one, i.e. when there is mutual attraction among the two species or groups of species. A value of zero indicates that the distributions of the two species are independent of each other. $S > 0$ indicates spatial segregation. The segregation index was calculated by Füldner and Gadow (1994) for six sample plots in a mixed deciduous Beech forest in Northern Germany, based on two groups of species (Beech and others) before and after a thinning. The results are presented in Table 5-2.

	Plot 1	Plot 2	Plot 3	Plot 4	Plot 5	Plot 6	Mean
Before	-0.105	0	0	-0.233	-0.154	+0.095	-0.066
After	-0.111	0	0	-0.438	+0.065	0	-0.080

Table 5-2. Index of segregation for two species groups (Beech and other species) in a mixed deciduous Beech forest in Northern Germany.

The index assumes zero values in several plots. Some of the index values are not at all affected by the thinning, which means that neither segregation nor repulsion between Beech and the other species (mainly Maple and Ash) nor any change of this condition has been observed. Mutual attraction is enhanced by the thinning in Plot 4 while the opposite is found in Plot 5.

An advantage of Pielou's index is the fact that it can be used to test whether an observed value is significantly different from zero. Unfortunately, its use is limited to first-neighbor constellations, which may be undesirable. Theoretically, a tree of species A can have up to 5 surrounding nearest neighbors of species B. Each B in turn may have an A as its nearest neighbor. Such a constellation, which may give an inflated impression of the degree of attraction, is shown in Fig. 5-4.

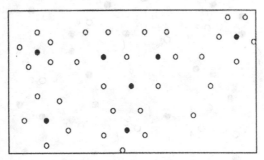

Figure 5-4. Hypothetical stand showing inflated attraction between two species.

The index of segregation is liable to be inflated either by a clustered or by a regular distribution of tree positions (Upton and Fingleton, 1990, p. 244). Another problem is the fact that Pielou's index in this form does not lend itself to evaluating species-specific mingling, which may be rather important information in certain forest types. We will attempt to solve this problem by considering not one, but n nearest neighbors of a given tree. *Mingling* of the i'th tree (M_i) is then defined as the relative proportion of neighboring trees of different species:

$$M_i = \frac{1}{n} \sum_{j=1}^{n} v_{ij}$$ 5-4

where v_{ij} is a discrete binary variable, which assumes the value 0, if the neighboring tree j is of the same species as the reference tree i, and the value 1, if it is not. Obviously, $0 \le M_i \le 1$.

The principle is illustrated in Fig. 5-5 using species-specific mingling. The three lines point to the three nearest neighbors of each of the 10 trees of species B. M_1 and M_2 are both equal to 1, the values of M_3 and M_4 each amount to 2/3, and so forth. The average value of the M_i's in the hypothetical example is 0.87, indicating that species B does not display repulsion in respect to A.

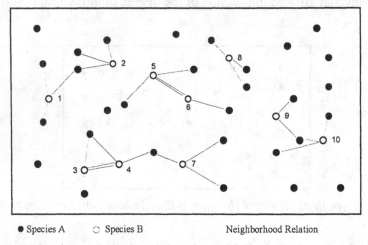

● Species A   ○ Species B   Neighborhood Relation

Figure 5-5. Hypothetical section of a forest with two tree species. Nearest-neighbor constellations are shown for calculating the species-specific mingling for B, using n=3.

The effect of a thinning on the average mingling in 6 sample plots is shown in Tab. 5-3, separately for beech and ash. The data are derived from a mixed deciduous beech forest in Northern Germany (according to Füldner and Gadow, 1994).

Beech

	Plot 1	Plot 2	Plot 3	Plot 4	Plot 5	Plot 6	Average
Before	0.059	0.149	0.492	0.299	0.372	0.449	0.276
After	0.092	0.175	0.456	0.262	0.354	0.409	0.270
Increase	+0.033	+0.026	-0.036	-0.037	-0.018	-0.040	-0.006

Ash

	Plot 1	Plot 2	Plot 3	Plot 4	Plot 5	Plot 6	Average
Before	no ash	0.667	0.500	0.844	0.833	0.533	0.638
After	no ash	0.722	0.569	0.897	0.867	0.540	0.683
Increase		+0.055	+0.069	+0.053	+0.034	+0.007	+0.045

Table 5-3. Species-specific mingling for beech and ash before and after a thinning in a mixed deciduous beech forest, based on the three nearest neighbors.

The values for beech are much lower than those for ash. This is partly the result of the high proportion of beech trees in the plots. The rather high values for ash in plots 2, 4 and 5 are an indication that the species does not form homogeneous groups, and that ash trees are mingled on a tree-to-tree basis. The thinning had a slightly positive effect on mingling in ash in all the plots where ash occurred, while the changes affecting beech were both positive and negative.

The average value of the M_i is a stand attribute indicating the degree of species mingling. It may be a useful parameter for management. More important for the analysis of stand structure, however, is the distribution of the M_i.

Aggregation

Aggregation is a spatial attribute which describes the degree of regularity in the distribution of tree positions. A number of variables have been used for assessing and evaluating aggregation (Smaltschinski, 1981; Wenk et al., 1990, p. 207 et sqq.). A rather popular one is the index R developed by Clark and Evans (1954), which represents the ratio between the observed and the expected average distance between neighboring trees:

$$R = \frac{\text{observed average distance between neighboring trees}}{\text{expected average distance between neighboring trees}} = \frac{\frac{1}{n}\sum_{i=1}^{n} r_i}{\frac{1}{2} \cdot \sqrt{\frac{10000}{N}}} \qquad 5\text{-}5$$

with r_i = the distance of the i'th tree to its nearest neighbor (m),

               N = the number of trees per hectare and n = the number of trees in the plot.

Equation (5-5) needs to be modified for small sample plots, using a method proposed by Donnelly (1978), who estimates the expected average with an empirical equation based on the plot area, the length of the plot perimeter and the number of trees within the plot.

R takes on a value of 1 if the distribution of tree positions is random and tends toward zero with increasing aggregation. Values greater than 1 indicate increasing regularity, the maximum possible value being 2.1491. The values of R, calculated for 6 sample plots in a mixed deciduous Beech forest in Northern Germany, are given in Tab. 5-4 (Füldner and Gadow, 1994).

	Plot 1	Plot 2	Plot 3	Plot 4	Plot 5	Plot 6	Mean
Before	1.12	1.35	1.32	1.03	1.17	1.17	1.19
After	1.09	1.27	1.17	1.10	1.14	1.22	1.16

Table 5-4. Values of the index of Clark and Evans modified with Donnelly's correction, calculated for 6 sample plots in a mixed deciduous Beech forest in Northern Germany, before and after a thinning.

Before the thinning, the values of R indicate on average a slight tendency towards regularity of tree positions, which is only marginally reduced by the thinning. There are differences in the individual plots, indicating a spatial variation of thinning effects on stand structure. In plots 2 and 3, for example, regularity was reduced considerably more than in the other plots.

Pommerening (1997) described spatial structure using the average distance of the i'th tree to its n nearest neighbors (A_i):

$$A_i = \frac{1}{n}\sum_{j=1}^{n} s_{ij} \qquad\qquad 5\text{-}6$$

with s_{ij}= distance from tree i to neighboring tree j (m).

The spatial structure is an important forest attribute because forests with the same frequency distribution may have completely different spatial structures. As an example, consider Fig. 5-6 showing three hypothetical forests, each one containing the same 47 trees. The forests are identical with regard to the numbers of trees per species and with regard to the diameter and height distributions. The only difference is in the spatial structure, in the way that tree positions and attributes are spatially arranged.

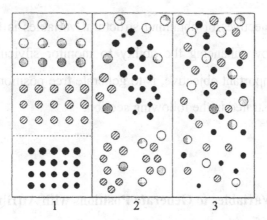

Figure 5-6. The same 47 trees spatially arranged in three different ways, resulting in very different spatial structures.

The distributions of the T_i- and M_i-values reveal the spatial structure. These distributions, corresponding to the three forests in Fig. 5-6, are shown in Fig. 5-7, considering the two nearest neighbors.

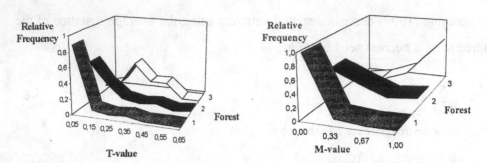

Figure 5-7. Distribution of T_i and M_i-values corresponding to the three forests in Fig. 5-6.

The interpretation of such structure graphs is straightforward. About 80 percent of the trees in the first stand, with the regular pattern, have zero mingling and very low differentiation. In the third stand, on the other hand, there is no tree with zero mingling, and 20% of the trees have differentiation equal to 0.5. Such an analysis is easily performed by a forester wishing to evaluate different thinning options.

One of the objectives of obtaining structural variables is to improve forest descriptions, thereby enhancing the ability to monitor silvicultural operations. Another, equally important objective, is to generate tree positions with attributes. This will be demonstrated in the following section.

Using Structural Variables to Generate Positions with Attributes

As mentioned before, a prerequisite to using a position-dependent individual tree growth model is the ability to *reproduce* a forest by generating tree positions with attributes for the entire forest. As it is impossible in practice to measure all the tree positions, the reproduction must be achieved with limited information. The information typically available to foresters involves distributions of neighborhood relations obtained during a routine forest inventory. A precondition to obtaining such information is the application of new inventory techniques, such as the structural group method[1].

Unfortunately, *point processes*, though appealing in theory, are limited to the description of spatial distributions (Tompo, 1986; Penttinen et al., 1994; Degenhardt, 1995). The most promising technique, known as *marked point processes*, has not yet been successfully applied for reproducing forest structures from samples. Thus, for the time being, simulation techniques, such as developed by Pretzsch (1994) and Lewandowski and Gadow (1997), seem to be the only practical alternative. The simulation generates a reproduction of a real forest, based on a sample of the structural variables described in the previous section.

A reproduction of a forest will be considered perfect, if each tree in the real stand has a counterpart in the reproduced one, with exactly the same distances to its 3 nearest neighbors, and if all the M_i-values and all T_i-values in the reproduction occur with the same frequencies as those in the real forest. The simulation consists of 4 separate phases, during which tree positions are generated and shifted, and position attributes exchanged, until the distributions of the structural variables of the reproduction are close enough to those of the real stand.

The initial positioning of the trees is done in phase 0. The positions may be allocated either at random or in some predefined manner. The denser the stand, the less important is the manner in which initial coordinates are generated. Phase 1 consists of 3 cyclic subphases: optimizing the distances to the first, the second and the third neighbor. The subphases are repeated until all the distances agree with those of the sample from the real forest.

---

[1] The sampling technique is known as the *structural group of 4*. Samples are taken on systematic grid points. The *group of 4* consists of the tree nearest to the sample point and its 3 nearest neighbors (Füldner, 1995). It is a simple and effective method for obtaining distributions of *mingling* and *differentiation*.

Fig. 5-8 shows a section of a fully-enumerated forest and its simulated counterpart, with the distances to the first neighbors. The numbers indicate the ranks of the trees in a list sorted according to distance.

The distance of tree No. 8 to its nearest neighbor, for example, is 3.20 m in the artificial forest, but only 2.67 m in the real forest. Other trees are closer together than they should be, e.g., 1 and 2. Trees are moved closer together or further away from their first neighbors, depending on the comparison of the original forest with the simulated one.

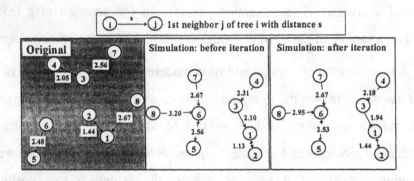

Figure 5-8. Distances to first neighbors in the real forest (shaded) and the simulated one (unshaded), before and after one iteration.

Fig. 5-8 shows the improvement after the iteration. Tree 8, for example, is located closer to its neighbor with 2.95 m. All the other values have been improved, except those of tree No. 3. The coordinates are fixed after completion of phase 1.

The purpose of phase 2 is to optimize the *mingling*. This is achieved by successively exchanging two trees of different species (swapping trees of the same species would not change the mingling value)[1]. The swap is retained if it

---

[1] A total enumeration would be a fairly hopeless endeavour. For example, in a forest containing 47 trees (with 20 trees of species 1, 15 of species 2 and 12 of species 3), there are

$$\binom{47}{20} \cdot \binom{27}{15} \cdot \binom{12}{12} = 1.697 * 10^{20}$$

different ways in which the species can be assigned to the available positions.

reduces the difference between the real and the simulated distributions of the mingling values, otherwise it is repealed. Phase 2 terminates when the difference between the two distributions of the mingling values cannot be further reduced.

The *diameter differentiation* is optimized in phase 3. Again, a pairwise swapping of trees of the *same species* is done with the aim of attaining a distribution of the T_i-values, which closely resembles that of the original forest.

The algorithm, which is described in detail by Lewandowski and Gadow (1997), produced satisfactory results using the structure variables *mingling* and *differentiation*. A methodical alternative is described by Pretzsch (1994), who was able to reproduce spatial structures based on available stand attributes describing the manner in which the different species mingle (e.g., in clusters, rows or groups).

The ability to reproduce forest structures from field samples is an important precondition if position-dependent individual tree growth models are to be used.

Competition Indices

In order to grow, a tree needs light, carbon dioxide, water and minerals. These elementary substances are converted into complex organic molecules as a result of very specific chemical reactions. A tree does far more than simply increase its volume as it grows. It differentiates and takes shape, forming a variety of cells, tissues and organs. It has the ability to respond and adjust to a wide range of changes in its environment. This ability is manifested in changing patterns of growth (Mitscherlich, 1971, 1975; Raven et al., 1987).

Many of the details of how these processes are regulated are not known. However, it is generally accepted that a primary factor influencing tree growth is spatial interaction. The interaction is not neccessarily antagonistic as implied by the term *competition*, but involves positive aspects of coexistence such as mutual adaptation, dependence or protection (Trepl, 1994). Clearly, competition is not the only basic relation between the elements of a biological system. Forests are highly integrated communities with the relations between the individual trees often being supportive. This aspect, however, is not part of competition modelling.

Single-tree spatial models use information about the distances and sizes of neighboring trees, and this effect is summarily designated as *competition* or *point density*[1]. Trees do not always compete for limited water, nutrients or light. Often, competition is simply a matter of physical obstruction or constriction by neighboring individuals, which are close enough to impede the expansion of a tree crown.

There are numerous measures designed to quantify tree competition, and the possibilities for inventing new ones are endless. Most of the existing competition indices incorporate the area potentially available to a tree, or the sizes, distances and directional distribution of neighboring trees, using either 2-dimensional or 3-dimensional spatial data (Clutter et al., 1983; Tomé and Burkhart,1989; Pukkala, 1989; Holmes and Reed 1991; Biging and Dobbertin, 1992, 1995; Vanclay, 1994, p. 58 et sqq.).

---

[1] A distinction must be made between *competition* and *density*. Competition refers to the spatial interaction between neighboring trees, while density is a population attribute, a measure of how completely a site is occupied by trees. The term *point density* is sometimes used in the North American literature as a synonym for *competition*.

Overlapping Influence Zones

Many studies of inter-tree competition involve the concept of a competition influence zone around each tree. The assumption is that the area, in horizontal projection, over which a tree competes for the resources of the site, can be represented by a circle. The radius of the circle is a function of tree size. The amount of competition experienced by a given tree is then assumed to be a function of the extent to which its influence zone overlaps those of neighboring trees (Staebler, 1951). Fig. 5-9 shows a hypothetical influence zone for each of 6 trees and the shaded areas of overlap for the reference tree i.

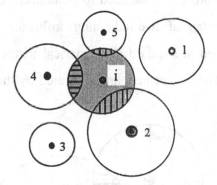

Figure 5-9. Influence zones of reference tree i and 5 potential competitors. Shading indicates area of overlap with competitors 2, 4 and 5.

Different functions of tree size for calculating the influence zone radius and different procedures for weighting the overlap areas with the relative sizes of the competitors, have led to a proliferation of indices. One of the prototypes was defined by Gerrard (1969) as follows[1]:

$$CI_i = \sum_{j=1}^{n} \left(\frac{a_{ij}}{A_i} \right)$$

                                                                              5-7

with a_{ij} = influence zone overlap between subject tree i and competitor j,
      A_i = influence zone of subject tree i.

---

[1] See applications by Newnham (1966), Opie (1968), Bella (1971), Arney (1973), Ek and Monserud (1974).

A reasonable approach for calculating the diameter of the influence zone is to use the potential crown width of a free-growing tree. This quantity is usually estimated from breast height diameter. An example is provided by Villarino Urtiaga and Riesco Muñoz (1997), who developed the following regression for estimating the crown width of a free-grown birch tree (*Betula celtiberica*) in Galicia, Spain:

$$\frac{dkra}{d} = 35.46 - 0.3229 \cdot d \qquad \text{5-8}$$

where dkra=diameter at base of crown [cm] and d=breast height diameter [cm].

One of the technical problems that need to be addressed involves the calculation of the area of overlap of two adjoining influence zones. The appropriate procedure is presented in standard mathematical textbooks. Consider two discs with radii r and R. The disc centers are separated by the distance s, as shown in the following graph:

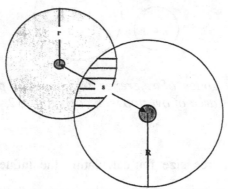

The shaded area of overlap may be calculated using the following equation, which is valid for r-R<s<r+R:

$$\text{Area}(s, r, R) = r^2 \left\{ \arccos\left(\frac{s^2 + r^2 - R^2}{2tr} \right) - \frac{s^2 + r^2 - R^2}{4t^2 r^2} \left[4s^2 r^2 - \left(s^2 + r^2 - R^2\right)^2 \right]^{\frac{1}{2}} \right\}$$

$$+ R^2 \left\{ \arccos\left(\frac{s^2 + R^2 - r^2}{2sR} \right) - \frac{s^2 + R^2 - r^2}{4s^2 R^2} \left[4s^2 R^2 - \left(s^2 + R^2 - r^2\right)^2 \right]^{\frac{1}{2}} \right\} \qquad \text{5-9}$$

Thus, Equation 5-9 is applicable in all cases, except those that are either irrelevant (there is no overlap at all between the two trees) or very easy to calculate (the influence zone of one tree is located entirely within the influence zone of the other).

The value of the competition index increases with increasing number of trees with large influence zones found in the immediate vicinity of the reference tree. The index assumes the value *zero*, if there is no overlap at all within the influence zone *i*. A value of *one* may be obtained with various constellations: e.g. the entire influence zone *i* is covered by part of the influence zone of one large neighbor or by several smaller neighboring zones. Obviously, the index may assume values greater than one. This happens when several tall trees are close enough, overtopping the reference tree.

As indicated before, an index is a measure designed to facilitate comparison between a given situation and one or more standard ones that provide interpretable reference points. The index of overlapping influence zones has only one interpretable reference point, the lower bound value zero. The upper bound value is not clearly defined, which may be seen as a disadvantage. Another problem is the size of the influence zone that is not defined *a priori*. A plausible approach is to set it equal to the area covered by the crown of a solitary tree, as demonstrated by Ek and Monserud (1974):

$$CI_i = \sum_{j=1}^{n} \left(\frac{a_{ij}}{A_i} \cdot \frac{S_j}{S_i} \right)$$

                                                                 5-10

with a_{ij} = influence zone overlap between subject tree i and competitor j,
        A_i = influence zone of subject tree i,
        S_j/S_i = size weighting factor (S=tree height multiplied by the potential crown radius of a solitary tree).

Distance-weighted Size Ratios

Another group of indices measuring point density and competition are the distance-weighted size ratios. These indices are based on the hypothesis that the competition effect exerted by a neighboring tree increases with increasing size and decreasing distance. A measure of tree size that is usually available is the tree diameter, and the index suggested by Hegyi (1974) can be used in many situations where there is uncertainty about the radius of the influence zone, i.e. where the potential cross-sectional crown area cannot be calculated:

$$HgCI_i = \sum_{i \neq j} \left(\frac{D_j}{D_i} \cdot \frac{1}{Dist_{ij}} \right)$$ 5-11

with D_i = breast height diameter of reference tree i,
               D_j = breast height diameter of competitor tree j (j ≠ i)
               $Dist_{ij}$ = distance between subject tree i and competitor j

Equation (5-11) defines a distance-weighted aggregate of the diameter ratios[1]. Again, only the lower bound value provides an interpretable reference point. Obviously, not all the trees in the forest are active competitors (Fig. 5-10). Thus, some rule must be developed to limit the search radius. Hegyi (1974) specified a competition circle with a radius of 10 ft (3.05m), counting all the trees within the circle as competitors. However, as the number of trees in the circle will gradually decline, the index value will automatically decrease with increasing age. Daniels (1976) defined a competitor as any tree that would be counted by a factor-10ft²/acre angle gauge centered at the reference tree, while Ford and Diggle (1981) included all individuals taller than a 45° angle from the top of the reference tree. Biging and Dobbertin (1992) included trees within a search radius, for which the angle from the crown base of the reference tree to the competitor's tip exceeds a specified threshold value. A number of such rules have been applied, one stating that a tree is a competitor, if the distance from tree i to tree j

is less than the sum of the two breast height diameters i and j divided by 8. None
of these rules appreciates the fact that several competitors may be lined up behind
each other, when viewed from the reference tree.

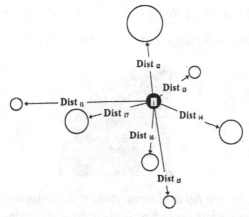

*Figure 5-10. Hypothetical example of a reference tree i (shaded) with 7 potential
competitors of different dimensions, located at various distances.*

The problem of defining *active* competitors was addressed by Lee and Gadow
(1997) in a different way. They suggested an iterative procedure for utilizing
sample plot data in a *Pinus densiflora* forest in Korea. Initially, all the trees
within a specified competition zone (CZ) are considered potential competitors of
a given reference tree. The competition zone (CZ) for each reference tree is a
circle with radius (CZR):

$$CZR = k \cdot \sqrt{\frac{10\,000}{N}}$$ 5-12

with k = a constant defining the radius of the competition zone (2<k<4) and N= number of
trees per ha.

Only the neighbor closest to the reference tree is considered an active competitor,
when several potential competitors are lined up behind one another (as viewed
from the direction of the reference tree). A potential competitor is said to be
positioned *behind* another, if it is located within a *competition elimination sector*

---

[1] See also applications by Lorimer (1983); Martin and Ek (1984).

defined by a *competition elimination angle*. Fig. 5-11 illustrates the procedure. The *competition elimination sector* is indicated by the dark shading. The nearest neighbor of the reference tree is No. 1, which is an active competitor. Numbers 2 and 3, located behind Number 1 in the *competition elimination sector* (defined by a 90° *competition elimination angle)* are not active competitors.

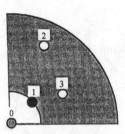

Figure 5-11. Reference tree (0) and three potential competitors (1,2,3). Tree No. 1 is the closest neighbor and identified as an active competitor. Trees 2 and 3, located behind Tree 1, are not active competitors.

Thus, each neighbor of a given reference tree within the competition zone is evaluated to establish whether it is an *active* or an *inactive* competitor.

An example illustrating the iterative procedure is presented in Fig. 5-12. The competition zone around the reference tree 0 contains 10 *potential* competitors. The nearest neighbor of the reference tree is No 1, an *active* competitor. Neighbors 2 and 3, located *behind* Number 1 and within the first *competition elimination sector* (defined by a 30° *competition elimination angle)*, are inactive competitors. The second *active* competitor, located outside the first *competition elimination sector* and facing the reference tree, is Number 8. Numbers 6 and 7 are inactive. Tree Number 4 is the next active competitor, and so on. The procedure terminates when all the active competitors have been identified[1].

---

[1] If two competitors are located *alongside* each other, facing the reference tree, they are both classified as *active*. The adjective *alongside* is true if the horizontal distance to the reference tree is within specified bounds of the distance to the already established active competitor and if the horizontal angle relative to the reference tree is greater than a given threshold angle (Lee and Gadow, 1997).

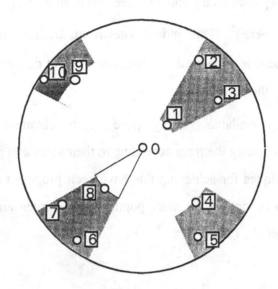

Figure 5-12. Hypothetical example of a competition zone around the reference tree 0 with 10 potential competitors, of which only 4 are active ones.

Partial correlation coefficients were calculated, depicting the strength of the relationship between the HEGYI-Index and current diameter increment. The results of the analysis indicate that in a given pine forest in the Kangwon Province, Korea, the competition-elimination angle should not exceed 30°, if the aim is to explain variations in diameter growth, based on the HEGYI-Index.

Available Growing Space

Measures for describing point density have been used for many years. Perhaps the earliest one is the *available growing space* index (*Wachsraumzahl*) applied by Seebach (1846)[1]. The *available growing space* index *w* is simply the observed crown diameter of a tree divided by its breast diameter. Alternatively, the *available growing space* has been calculated as the ratio between the crown

---

[1] Sterba (1991, p. 41); see also the reference by Spiecker (1994) to applications by Kraft in 1884.

projection area $\pi/4(\text{crown width})^2$ and the tree basal area $\pi/4(\text{tree diameter})^2$, i.e. $(\text{crown width}/\text{tree diameter})^2$. This index, known in German literature as the *Ausladungsverhältnis*, may be used as a measure of point density. The smaller its value, the greater is the point density.

The potentially available growing space may be obtained by apportioning the total forest area among the trees according to their sizes and positions (Fig. 5-13). Various techniques for achieving this have been proposed and the *Voronoi* method appears to be one of the more popular ones (Green and Sibson, 1977; Lee, 1980; Nance et al., 1988).

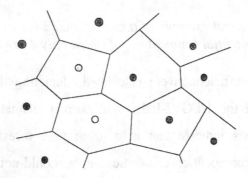

Figure 5-13. Subdivision of a forest area for assigning potentially available growing spaces to individual trees.

A *Voronoi* polygon contains the set of all points which are positioned closer to the *polygon tree* than to any other tree. Computationally, the construction of Voronoi polygons involves two tasks: a) locating a set of nearest neighbors for each tree, and b) defining the edges of each polygon. In determining the location of each edge, connecting lines between the reference tree and each of its neighbors are established. The intersecting perpendiculars form the edges of the polygons. In the case of unweighted tessellations, the perpendiculars are placed exactly halfway between the two neighbors. In weighted tessellations, the perpendicular intersections are displaced further away from the larger tree and closer to the smaller one by an amount proportional to their size differences. Standard algorithms are available to calculate the polygons, ensuring that they do not overlap and that no unrealistic gaps remain (Lewandowski and Pommerening, 1996).

Pukkala (1989) describes a simulation study based on a mapped stand of *Pinus silvestris* with an application of *ecological field theory* (Wu et al., 1985). According to this concept, each tree reduces the availability of radiation, water and nutrients by its roots, crown and stem. Each of these morphological entities generates a *field* of resource use that can be described mathematically. It is possible to determine the amount of an essential resource which is accessible by a given tree, based on the area potentially available to it.

The areas potentially available are not static, but constantly change as the trees grow, or as neighboring trees are removed during a thinning operation, thus creating additional growing space. Rennols and Smith (1993) therefore suggest a dynamic extension of the concept of partitioning of resource spaces by introducing the concept of *evolving overlap* of the zones of influence.

Shading and Constriction

The most common forms of physical impediment within a group of trees interacting with each other, are constriction of the growing space and shading from above. Constriction by neighboring trees limits the expansion of a tree crown (Assmann, 1953; Fig. 5-14), while shading reduces the amount of light required for photoproduction.

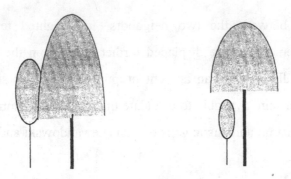

Figure 5-14. Two common forms of impediment: physical constriction of the growing space (left) and shading from above (right).

These types of interaction between neighboring trees can be modelled using a three-dimensional spatial approach. Based on the known tree coordinates, heights and crown dimensions, it is possible to define a three-dimensional spatial matrix with cells that have attributes. For example, using a cell size of $1m^3$, a forest of 1 hectare with a maximum tree height of 25m would consist of 250000 cells. Associated with each cell is a vector of attributes (*tree number, species, stem, crown*, etc).

The amount of shading may be assessed using a cone of light for each tree at 70 percent of its total height (Sloboda and Pfreundt, 1989). The greater the number of cells containing foreign crown biomass in the cone of light above a given tree, the heavier is the amount of shading. Thus, it is simply a matter of defining the coordinates of the cell midpoints within the cone, and establishing the frequency of hits (Fig. 5-15).

The shading index used by Pretzsch (1992, p. 147, 249) is derived on the basis of the number of cone cells containing biomass of neighboring crowns:

$$w = \sum_{x=i}^{imax} \sum_{y=j}^{jmax} \sum_{z=k}^{kmax} \frac{b_{x,y,z}}{d^2_{x,y,z}} \qquad \text{5-13}$$

with x,y,z = coordinates of cell midpoints in the cone of light;

$$b_{x,y,z} = \begin{cases} 0, \text{ if cell}(x,y,z) \text{ hits nothing} \\ \\ 1, 2 \text{ or } 3, \text{ if cell}(x,y,z) \text{ hits } 1, 2 \text{ or } 3 \text{ trees} \end{cases}$$

$d_{x,y,z}$ = distance between midpoint of cell x,y,z and the lower point of the cone of light.

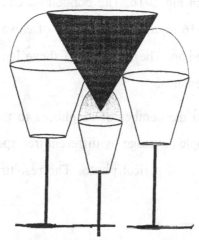

Figure 5-15. Cone-of-light model for quantifying the amount of shading for a tree at 70 percent of its total height.

The side length of the cells within the cone is set relative to the tree height (e.g. one-twentieth of the tree height) to obtain values of ϖ, which are independent of tree size. Computer time can be reduced by not evaluating the entire cone, but only every n'th layer of cells.

In a three-dimensional modelling approach, the amount of shading may be measured in a variety of ways. Biber (1996), for example, developed a method for simulating hemispherical *fisheye*-photos in mixed Beech/Spruce forests. The simulated photo is "taken" from the center of the crown at a specified reference height (e.g., in Beech trees, a distance from the tip of 1/5 of the crown length). The *sky-view* data are based on the known tree positions, crown lengths and crown widths of the neighboring trees (Fig. 5-16).

The amount of shading is assessed on the basis of the photo coverage, separately for various sectors of the discretized hemispere (c.f. Anderson, 1964; Courbaud, 1995; Biber, 1996). In this way, it is possible to evaluate directional influences and to compare the effects of a shading from above with one from the sides. This is illustrated in Fig. 5-16. The concentric circles in the graph on the right-hand side of Fig. 5-16 represent the angles of elevation of the objects that are the source of the shading. The imaginary fisheye lens, directed towards the zenith, is located at the center.

Biber (1996) limited the number of neighbors to the 50 nearest trees and evaluated simple or multiple coverage by different tree species within the various sectors of the simulated hemisperical photo. The resulting competition index is based on this information.

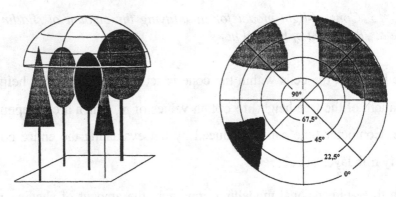

Figure 5-16. Generating a simulated hemispherical fisheye-photo with a 180° sky view in a mixed Beech/Spruce forest (left) and evaluating the results for different sectors according to different directions and angles (right).

The *constriction* of the crown of a tree by its neighbors can be measured in a similar fashion by counting the number of overlapping cells at a specified height. Fig. 5-17 demonstrates the technique applied by Pretzsch (1992). The actual and the potential cross-sectional crown area of the reference tree A (the *reference disc*), taken at the height of maximum crown width, is indicated by dark and light shading, respectively. 100 grid points are assigned to the reference disc, irrespective of its area F. Thus, the side length of each square grid cell is equal to $\sqrt{F/100}$.

The index of *constriction* used by Pretzsch (1992, p. 248) is calculated by counting the number of grid cells within the areas of overlap:

$$\varepsilon = \sum_{k=1}^{100} t_k \cdot 0.01$$

5-14

with k=number of the grid cell and t_k=contribution of grid point k to the total constriction ε.

and where $t_k = \begin{cases} 0, \text{ if cell k hits nothing} \\ \\ 1,2 \text{ or } 3, \text{ if cell k hits 1, 2 or 3 trees} \end{cases}$

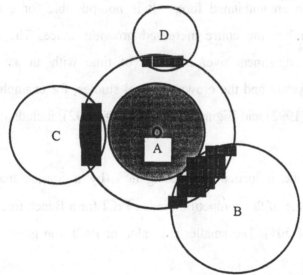

Figure 5-17. The constriction of the crown of tree A by three neighbors B,C and D is measured by counting the number of overlapping cells. Dark shading indicates the actual, light shading the potential cross-sectional crown area, taken at the height of maximum crown width.

Thus, considering shading and constriction simultaneously, the following competition index was calculated for a tree numbered i by Pretzsch (1992) in a mixed Spruce/Beech forest, based on available 3-dimensional spatial data:

$$\Pr CI_i = 0.89 \cdot e^{0.25 \cdot \varepsilon_i - 0.14 \cdot \varepsilon_i^2} \cdot e^{-0.44 \cdot \omega_i^{Fi} - 0.21 \cdot \omega_i^{Bu}} \qquad\qquad 5\text{-}15$$

with ε_i = index of constriction calculated for tree i;
ω_i^{Sp}, ω_i^{Be} = index of shading caused by neighboring Spruce (*Picea abies*) or Beech (Fagus silvatica) trees, respectively, for tree i.

Few studies dealing with competition effects include the reaction of tree growth to thinnings. The removal of a neighboring tree causes abrupt modifications of the point density and an immediate reduction of the competition index value. Intuitively, the resulting condition should, at least for some time, be unlike any comparable condition in an unthinned forest. It is not possible for a tree to instantly occupy and utilize the entire increased growing space. The normal response is a gradual adjustment over a period of time with an associated expansion of the root system and the crown. Several studies, for example those conducted by Pretzsch (1992) and Biging and Dobbertin (1992), include thinning effects.

Fig. 5-18 shows the influence of shading ω^{Sp} (by a Spruce tree) and constriction ε on the value of the competition index PrCI for a Beech tree with a relative crown length of 50%. The smaller the value of PrCI, the greater is the growth reduction effect.

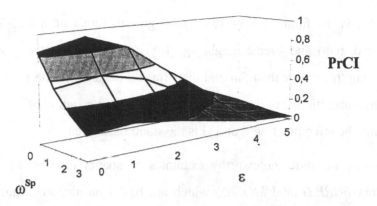

Figure 5-18. Value of competition index PrCI for different values of ε and ω^{Sp}.
The response refers to a Beech tree with a relative crown length of 50%.
The smaller the value of PrCI, the greater is the growth reduction effect.

Fig 5-18 shows that the development of the tree is impeded by increased shading, but to an even greater extent by increased constriction. Maximum growth occurs at zero shading. The growth-reducing effect of constriction becomes serious if the value of ε is greater than 2 (more than 20 of the 100 grid cells within the potential cross-sectional area of the crown are encountered in an area of overlap).

Spatial Growth Models

A number of spatial growth models have been developed, all of which are based on some measure of competition requiring information about tree positions and associated tree attributes. The following is an example of a spatial growth model developed by Pukkala (1989a) for predicting the basal area growth of individual Scots pine trees in Finland:

$$\ln(\Delta g_i) = -3.093 - 0.1815 \cdot d_i + 2.052 \cdot \sqrt{d_i} + 38.85 \cdot [1/(t_i + 10)]$$
$$- 0.009561 \cdot \sum_{j \neq i} (d_j / dist_{ij}) + 1.043 \cdot (d_i / Dg) - 0.01173 \cdot G$$

5-16

where Δg_i is the future 5-year basal area growth (cm²) of a tree with breast diameter d_i (cm) and breast height age t_i years; d_j (cm) is the diameter of a neighboring tree closer than 5m and $dist_{ij}$ (m) its distance to the reference tree i; Dg is the quadratic mean diameter of the trees within a radius of 5m from tree i (including the reference tree i) and G is the stand basal area.

Among the most noteworthy examples of spatial models are the growth simulators *FOREST* and *MOSES*, which are based on similar assumptions and principles, and the three-dimensional model *SILVA*. In principle, the use of these models is not limited to a particular type of forest. They require empirical parameter estimates and real or assumed tree data, and one of the possible applications is to provide some insight of growth processes that less detailed models cannot give.

FOREST, WASIM and *MOSES*

FOREST, one of the first position-dependent growth simulators, was conceived and developed by Ek and Monserud (1974). They applied the so-called *potential-modifier* approach. The assumption is that maximum or potential growth derived from observations of solitary trees is reduced by competition. The effect of competition is modelled using growth modifiers, and one of the noteworthy contributions of the *FOREST* concept is the simultaneous use of modifiers which embody past and present competition:

$$\Delta W = \Delta W_{pot} \cdot CR^b \cdot OVS \qquad\qquad 5\text{-}17$$

with　　　　ΔW　= estimated change of size variable W;
　　　　ΔW_{pot} = potential change of size variable W;
　　　　　CR　= crown ratio (crown length/tree height);
　　　　　　b　= species-specific parameter;
　　　　OVS　= overstocking multiplier.

It is assumed that the amount of *historical* stress, the competition that a tree was exposed to in the past, is represented by the present crown ratio. The smaller the crown ratio, the greater the historical stress, and the greater the negative impact on present performance.

Present competition is described by the *overstocking multiplier* (OVS), which is a function of the area of overlapping influence zones:

$$OVS = \left(1 - e^{\frac{1}{\beta \cdot CIA}}\right)^{\alpha} \qquad\qquad 5\text{-}18$$

with   CIA = adjusted competition index: $CIA = 1 + 0.1 \cdot CI \cdot \left(1 - \beta_1^{h+1}\right) \cdot \left(1 + \beta_2 \cdot \Delta CI\right)^{-\beta_3}$;

           CI = competition index according to Bella (1971): $CI_i = \sum\limits_{j=1}^{n} \left(\dfrac{a_{ij}}{A_i} \cdot \dfrac{S_j}{S_i}\right)$;

          a_{ij} = influence zone overlap between subject tree i and competitor j,

          A_i = influence zone of subject tree i,

        S_j/S_i = size-weighting factor (S=tree height multiplied by the potential crown radius of a solitary tree),

       ΔCI = change of CI caused by thinning,

         h = tree height,

$\beta, \alpha, \beta_1, \beta_2, \beta_3$ = species-specific parameters.

The equation for calculating CIA includes a term for predicting thinning effects, which is an essential element of spatial growth models.

The conceptual design of *FOREST* is based on simple logic. This explains its intuitive appeal, and it is not surprising that a number of subsequent models have been structured in a similar fashion.

One of the models based on the conceptual design of *FOREST* is the growth simulator *WASIM* for mixed forests of Spruce (*Picea abies*) and Scots pine (*Pinus silvestris*) in Austria (Sterba, 1990). A simplified version of the general system structure, which is used for diameter and height growth of the two species, is presented in Fig. 5-19.

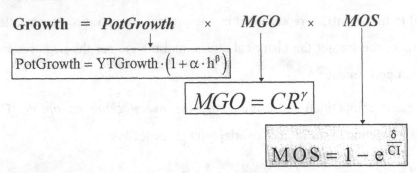

Figure 5-19. Simplified system structure of the growth simulator WASIM for mixed forests of Spruce and Scots pine (Sterba, 1990)[1].

The diameter or height growth of a tree is equal to the potential growth (*PotGrowth*) derived from a yield table (*YTGrowth*), multiplied by two reduction factors. The variable *MGO*, a measure of the historical competition, is a function of the crown ratio *CR* (the crown length expressed as a proportion of tree height *h*). The variable *MOS* describes the current competition based on a competition index *CI*. The parameter values for Spruce and Pine diameter and height growth are listed in Tab. 5-5.

Species	Type	α	β	γ	δ
Pine	d	0	0	0	-16.63
Pine	h	2.16 E-16	10.74	0.2903	-2.97
Spruce	d	1.16 E-12	8.54	0.4001	-4.81
Spruce	h	9.72 E-15	10.25	0.9368	-1.06

Table 5-5. Parameter values referring to the model PROGNAUS presented in Fig. 5-19 (d=diameter, h=height).

Example: Consider a Spruce tree with a total height of 20m. The *YTGrowth* for breast height diameter is 5mm/year. A few estimated diameter growth rates for various combinations of crown ratio and competition index values are listed in Tab. 5-6.

---

[1] refer also to the distance-independent model *PROGNAUS* which is using the *BAL*-index of competition (Sterba and Monserud, 1997).

CR	CI	Growth (mm/year)
0,6	1,5	1,9
0,6	0,5	3,3
0,6	1,0	2,5
0,3	1,0	1,3

Table 5-6. Estimated diameter growth rates for various combinations of crown ratio and competition index values.

Another derivative of *FOREST* and a successor to *WASIM* is the growth simulator *MOSES*, developed by Hasenauer et al. (1995). The three main elements of *MOSES* are systems for estimating height growth, diameter growth and the dynamic change of the height up to the base of crown.

Height growth is estimated from the potential yield table height growth modified by two reduction factors. Again, the crown ratio CR is a measure of historical competition, while the overlapping influence zones competition index CI describes current competition. Thinning effects are not included in Eq. 5-19:

$$\Delta h = \Delta h_{pot} \cdot CR^{\alpha} \cdot \left(1 - e^{\frac{-\beta}{CI}} \right)$$

5-19

with α=0.0845 and β=−6.158 for Beech and α=0.241 and β=−3.953 for Spruce. CI is Bella's index and CR is the crown ratio.

The estimation of the potential growth is a rather complicated process. Three different yield tables are used for deriving this important quantity. The choice of yield table depends on the forest type. The tables compiled by Assmann and Franz (1965) are used for Beech/Spruce forests. Kennel's (1972) table applies in pure Beech stands and Marschall's (1975) table in Spruce/Pine forests. A correction factor is used to derive the potential growth of individual trees from the average yield table growth of dominant trees in a stand with a given site index.

The *MOSES* diameter model has exactly the same form as the height model (Eq. 5-20), but the derivation of potential diameter growth is different. As data from free-grown solitary trees were not available, the potential diameter growth had to be estimated using the following equation:

$$\Delta d_{pot} = \frac{1}{0.529 \cdot H_2^{(0.589/-0.605)}} - \frac{1}{0.529 \cdot H_1^{(0.589/-0.605)}}$$ 5-20

with H=stand dominant height; the parameter values are applicable to Beech trees.

The third important element of *MOSES* is the change of the height to base of crown (HLC). This is necessary for estimating the crown ratio as an indicator of historical competition. The following crown model predicts this quantity from the tree diameter, thus avoiding the use of age as an independent variable:

$$\Delta HLC = b_0 \cdot h^{b_1} \cdot e^{\left(b_2 \cdot CR^{0.5} + b_3/CI + b_4 \cdot d\right)}$$ 5-21

with   HLC=height to base of crown, h=tree height, d=tree diameter,
CR=crown ratio, CI=Bella's competition index.

The parameter values for Beech are: b_0 =0.000014, b_1 =2.51, b_2 =6.08, b_3=−0.446, b_4 =−0.014. An example of a 5-year projection of the diameters, total heights and heights to crown base of 12 Beech trees and 3 Spruce trees with coordinates, representing a small section of a forest, is presented in Tab. 5-7.

No	species	initially observed tree data					after 5 years		
		d	h	HLC	X	Y	d	h	HLC
1	*Beech*	12.1	14.2	4.7	12.50	2.10	16.83	17.81	5.70
2	*Beech*	12.8	14.2	6.5	9.40	3.10	15.61	17.34	7.19
3	*Beech*	5.7	9.2	4.6	5.20	3.30	8.43	12.59	4.82
4	*Beech*	9.8	14.2	7.6	6.20	3.70	10.58	16.10	8.17
5	*Beech*	12.4	14.2	8.2	2.90	4.20	13.51	16.40	8.65
6	*Spruce*	12.8	12.5	8.4	12.40	4.30	13.93	14.25	8.99
7	*Beech*	6.3	10.0	3.5	5.90	4.70	7.50	12.47	4.02
8	*Beech*	8.8	12.7	6.5	5.40	4.80	9.37	14.39	6.99
9	*Beech*	5.5	9.9	6.2	5.10	5.00	6.51	12.14	6.36
10	*Beech*	4.4	8.2	5.9	14.00	5.50	10.63	12.18	5.94
11	*Beech*	6.2	11.1	8.3	4.90	5.60	6.82	12.85	8.41
12	*Spruce*	19.7	15.6	9.2	8.30	5.80	19.97	16.48	9.88
13	*Spruce*	19.4	14.7	7.3	10.60	5.80	20.49	16.33	8.18
14	*Beech*	9.3	16.1	8.0	13.10	5.80	11.04	18.65	8.89
15	*Beech*	9.7	14.0	5.4	6.00	6.10	10.01	15.27	6.45

Table 5-7. Initially observed and projected values of diameters (d), heights (h), heights to base of crown (HLC) and tree coordinates (X, Y) for 15 trees in a section of a mixed Spruce/Beech forest, using the relevant MOSES equations (Albert, 1997).

SILVA

Another growth simulator for individual trees showing some resemblance to the *FOREST* structure is the *SILVA* model which utilizes three-dimensional spatial data and was developed by Pretzsch (1992) for mixed Spruce/Beech forests in Germany.

The height growth of a Beech tree may be estimated using the following equation, which exhibits the same basic structure as the *FOREST* growth models with two multipliers, one for historical, the other for current competition:

$$\Delta h = \Delta h_{pot} \cdot CR^{0.088} \cdot PrCI \qquad\qquad 5\text{-}22$$

with $\quad \Delta h_{pot}$ = potential height growth; CR = crown ratio,

$\qquad$ PrCI = competition index based on *shading* and *constriction* (section 4.2).

The potential height growth of individual trees is estimated from stand age using a parabola with log-transformed variables. The following equation, for example, applies to a Spruce tree growing in a pure Spruce forest:

$$h_{pot} = 1.1304 \cdot e^{-9.14+5.43\cdot\ln(t)-0.574\cdot\ln^2(t)} \qquad\qquad 5\text{-}23$$

Diameter growth is derived in *SILVA* from basal area growth using an allometric relationship between the basal area increment, the initial basal area, crown volume and crown surface area, shading and constriction. The important crown parameters are calculated using a specific crown model developed by Pretzsch (1992, p.112 et sq.) for Beech (*Fagus sylvatica*), Spruce (*Picea abies*)and fir (*Abies alba*).

The crown is subdivided into two parts. The profile of the upper part is modelled using an exponential function, the profile of the lower shaded section is described by a linear one. Idealized forms are being used for the different species:

a simple cone for Spruce, a cubic paraboloid for Beech and a quadratic paraboloid for fir. The general form of the shaded part of the crown is the same for all species, namely a section of a cone.

Fig. 5-20 (left) shows a simplified model of the crown of a Beech tree using some descriptive parameters developed by Burger (1939): the upper part exposed to light, the shaded part, the maximum crown diameter, the crown length and the diameter at the crown base. The corresponding diagram on the right includes the variables required for modelling the two crown profiles (the lengths of the upper and the shaded parts, the maximum radius and the radius at the base) and the equations for the upper and the shaded parts.

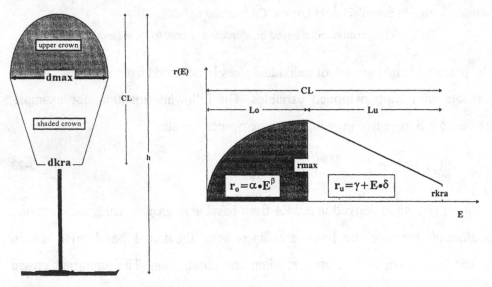

Figure 5-20. Model of the crown of a Beech tree. Left: dmax=maximum crown diameter; dkra=diameter at base of crown and CL=length of crown. Right: r(E)=crown radius (m) at distance E m from the tip; ro, ru = crown radius in the upper and shaded parts, respectively; rmax=maximum crown radius; rkra= radius at crown base; α, β, γ, δ=species-specific parameters.

The tree-specific (α,γ,δ) and species-specific (β) parameters of the crown model are listed in Tab. 5-8, for Spruce, Beech and Fir. The crown model provides the essential basis for calculating shading and constriction.

Species	Upper Crown			Shaded Crown		
	α	Lo	β	γ	δ	rkra
Spruce	r_{max}/Lo	$CL \cdot 0.66$	1.00	$r_{max}-\delta\cdot Lo$	$\dfrac{r_{kra}-r_{max}}{CL-Lo}$	$r_{max}\cdot 0.50$
Beech	$r_{max}/\sqrt[3]{Lo}$	$CL \cdot 0.40$	0.33	$r_{max}-\delta\cdot Lo$	$\dfrac{r_{kra}-r_{max}}{CL-Lo}$	$r_{max}\cdot 0.33$
Fir	$r_{max}/\sqrt{Lo}$	$CL \cdot 0.50$	0.50	$r_{max}-\delta\cdot Lo$	$\dfrac{r_{kra}-r_{max}}{CL-Lo}$	$r_{max}\cdot 0.50$

Table 5-8. Parameters of the crown model developed by Pretzsch (1992) for three different species.

The crown volume and crown surface area of a tree are needed for calculating its basal area growth. These quantities may be obtained on the basis of the known or assumed crown length and crown width. The equations for approximating crown volume *CV* and crown surface area *CA* are:

$$CV = \frac{dkra^2}{4} \cdot \pi \cdot CL \cdot kf \qquad\qquad 5\text{-}24$$

$$CA = a_0 \cdot dkra \cdot CL + a_1 \cdot dkra^{a_2} + a_3 \cdot CL^{a_4} \qquad\qquad 5\text{-}25$$

with CL = crown length (m) and dkra = crown width (m). The parameter values for Beech are, for example: kf=0.525; $a_0 = 2.43$; $a_1 = -0.172$; $a_2 = 0.576$; $a_3 = -0.038$; $a_4 = 2.087$.

Such relatively simple crown models can be used for generating three-dimensional forest structures and advanced competition indices such as those representing the amount of *shading* and *constriction*. This is possible, provided that relevant data are available from forest inventories, such as height to base of crown and crown width, or that these data can be estimated with reasonable accuracy.

Spatial data based on measured tree coordinates are hardly ever available in practice and methods to generate tree positions with attributes from forest samples are only beginning to emerge. The use of individual tree spatial models for forest planning and forecasting is therefore often considered an unrealistic alternative. It has been shown, however, that these models are capable of producing more realistic predictions of tree growth than simple distribution models, and that they can be used to explain variations in growth resulting from variations in forest structure (Pukkala, 1989; Pretzsch, 1994). Further development of these important tools is thus essential, not only for mixed, uneven-aged forests, but also for plantation forests with regular spatial distributions, which can often be reconstructed without difficulty.

Spatial Thinning Models

Forest development is the result of natural growth and modifications caused by silvicultural operations and other disturbances. Most important in a managed forest are the thinning operations, which are often carried out at regular intervals. Thus a model of forest development would be incomplete without a thinning component. Thinnings may be modelled either by imitating the way in which a forester normally selects trees for removal (*imitation*) or by prescribing thinning rules which must be obeyed, defining the specific manner of selection (*prescription*).

Imitation

Thinning decisions are normative, intrinsically fuzzy and influenced by a variety of criteria (Kahn, 1994). Thus, the ability to mimic the decisions that have led to the removal of individual trees will provide a better understanding of a thinning process (Daume, 1995). The fuzzy character of a thinning is exemplified by vague expressions (*high* thinning, *low* thinning), by the personalized nature of the decisions and by the fact that exact measurements of tree attributes are usually not available. Whereas a single attribute, such as the species, may be sufficient reason for *not* removing a tree, multiple attributes are required for explaining a positive selection for removal. Various combinations of *thinning-relevant* attributes may be used to quantify removal probabilities, depending on the kind of forest and the specific objectives of the thinning operation. After assigning these quantities, trees are sorted and those with the highest probabilities are removed.

Typical thinning-relevant attributes in a non-spatial situation are the tree species and variables defining the relative size or the economic value of a tree. A spatial model permits a much greater choice of criteria, such as the structural attributes of specific cells within a forest, or the point density in the immediate vicinity of a given tree. Tab. 5-9 shows the relative frequencies of the common distribution of the structural variables *mingling* (M) and *diameter differentiation* (T), both based on the three nearest neighbors[1] for a mixed stand of Beech and Spruce in Southern Germany. The stand was thinned in 1987 and the before-thinning (*total*) and removed data are shown separately for Beech and Spruce. The shading in the cells increases with increasing relative frequency.

Spruce					Beech				
Total		*M*			Total		*M*		
	0	0.33	0.67	1			0	0.33	0.67	1
	0 - 0.3	0.00	0.17	0.15	0.00		0 - 0.3	0.19	0.08	0.04	0.00
T 0.3-0.5	0.04	0.13	0.00	0.00	*T3* 0.3-0.5	0.07	0.07	0.04	0.01
0.5-0.7	0.00	0.02	0.09	0.02	0.5-0.7	0.04	0.06	0.07	0.01
0.7-1.0	0.00	0.06	0.11	0.06	0.7-1.0	0.13	0.07	0.08	0.03
Removed		*M*			Removed		*M*		
	0	0.33	0.67	1			0	0.33	0.67	1
	0 - 0.3	0.00	0.25	0.00	0.00		0 - 0.3	0.20	0.20	0.00	0.00
T 0.3-0.5	0.00	0.75	0.00	0.00	*T* 0.3-0.5	0.20	0.20	0.20	0.00
0.5-0.7	0.00	0.00	0.00	0.00	0.5-0.7	0.00	0.00	0.00	0.00
0.7-1.0	0.00	0.00	0.00	0.00	0.7-1.0	0.00	0.00	0.00	0.00

Table 5-9. Relative frequencies of the distributions of mingling and differentiation for the total population before thinning, and for the removed trees.

---

[1] See details about these variables at the beginning of Chapter 5.

Removal preferences may be calculated using the data in Tab. 5-9. Before the thinning, only 13 percent of the Spruce trees belonged to the mingling class [0.33][1] and the differentiation class [0.3-0.5]. However, 75 percent of all the trees in this class, having one Beech tree as one of the three nearest neighbors, were removed during the thinning. The selection preference in this class, calculated as the ratio of removed to total cell values, is very high.

Similar interpretations may be developed for the Beech trees. For example, 40 percent of all the removed Beech trees are found in pure clusters containing only Beech. In another 40 percent, two Beech trees and one Spruce tree are among the three nearest neighbors. The corresponding *total* cell values are not particularly high. Again, this points to a high selection preference in the two mingling classes. The preference is also abnormally high in the lower two differentiation classes, indicating preferred removal of trees surrounded by neighbors of equal or approximately equal size.

Thus, for the purpose of *imitating* a particular thinning operation, we may define a preference index as follows (c.f. Daume, 1995):

$$PR_{ij} = \frac{\text{removed } p_{ij}}{\text{before } p_{ij}} \qquad\qquad 5\text{-}26$$

with PR_{ij} = selection preference in structure class ij (i=1..m, j=1..n);
         removed p_{ij} = proportion of trees removed in structure class ij;
         before p_{ij} = proportion of trees in structure class ij, before the thinning.

The diagram in Fig. 5-21 shows the distribution of the PR_{ij} for the Spruce trees, derived from the data in Tab. 5-9. The PR_{ij} values are obtained by dividing the cell values in the *removed* class by the corresponding cell values in the *before thinning* class.

---

[1] Note that with 3 neighbors, the variable M can assume 4 possible values.

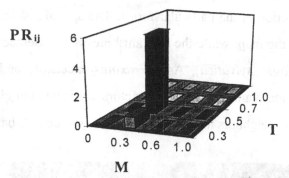

Figure 5-21. Distribution of the index of removal preference PR$_{ij}$, for the Spruce trees, based on the data in Tab. 5-9.

For example, the high value of 5.77 in the M-class 0.3 (precisely 0.33) and the T-class 0.3 (actually 0.3-0.5) was obtained by calculating the ratio 0.75/0.13. In this particular forest, the structure variables *mingling* and *differentiation* obviously provide an excellent basis for modelling the selection of trees for removal. The suitability of these variables is explained by the fact that they implicitly describe the immediate neighborhood constellation of each tree in a manner which must be highly relevant to the selection process.

The individual decisions involving selection of trees for removal are taken in *thinning cells*, i.e. areas which are sufficiently small for judging the effect of a particular removal on the remaining survivors, and large enough so that at least one tree is a potential candidate for removal. In practice, the forester who selects trees for removal moves from cell to cell, taking individual and independent decisions in each one. Consequently, it is plausible to use structure variables for calculating selection preferences. However, the variables should not only involve one, but several neighbors. The number of neighbors should be appropriate to the desired cell size, and the size of the cell should resemble the size that is used by the forester in the field.

Daume (1995) conducted an analysis of a thinning operation in a section of a 60-year old mixed Beech forest and found that the structural attributes of a tree are not sufficient for quantifying selection probabilities. Fig. 5-23 presents a

30x30m square section of the particular forest. The spatial distribution of the 32 trees is shown on the map, while the tree attributes are presented in the list: the species Beech (*Fagus sylvatica*), Ash (*Fraxinus excelsior*) and Maple (*Acer platanoides*), thé tree diameters and the structural variables *mingling* (*M*), based on the three nearest neighbors and *diameter differentiation* (*T*), based on the first neighbor.

Figure 5-22. Uneven-aged mixed forest in Germany, managed in the selection system

The trees numbered 13, 23 und 28 carry identical M- and T-values. However, only number 23 is removed. A similar observation can be made concerning the trees numbered 4 and 15, which carry identical attributes. Only number 15 is removed. Thus, knowledge about the structural attributes is obviously not sufficient for modelling a thinning, and additional spatial information is required. The objective is to predict removal decisions for specific trees within the context of a *thinning cell*. This approach imitates the way in which a forester marks trees for removal, moving from cell to cell and taking independent decisions, which are influenced only by the point density and structure in each particular cell.

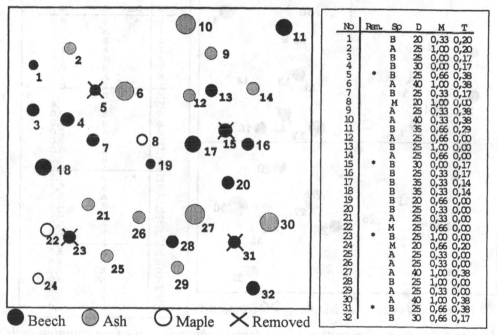

No	Rem.	Sp	D	M	T
1		B	20	0,33	0,20
2		A	25	1,00	0,20
3		B	25	0,00	0,17
4		B	30	0,00	0,17
5	*	B	25	0,66	0,38
6		A	40	1,00	0,38
7		B	25	0,33	0,17
8		M	20	1,00	0,00
9		A	25	0,33	0,38
10		A	40	0,33	0,38
11		B	35	0,66	0,29
12		A	25	0,66	0,00
13		B	25	1,00	0,00
14		A	25	0,66	0,00
15	*	B	30	0,00	0,17
16		B	25	0,33	0,17
17		B	35	0,33	0,14
18		B	35	0,33	0,14
19		B	20	0,66	0,00
20		B	25	0,33	0,00
21		A	25	0,33	0,00
22		M	25	0,66	0,00
23	*	B	25	1,00	0,00
24		M	20	0,66	0,20
25		A	25	0,33	0,00
26		A	25	0,33	0,00
27		A	40	1,00	0,38
28		B	25	1,00	0,00
29		A	25	0,33	0,00
30		A	40	1,00	0,38
31	*	B	25	0,66	0,38
32		B	30	0,66	0,17

● Beech   ◑ Ash   ○ Maple   ✕ Removed

Figure 5-23. Example of a thinning in a 30 x 30m section of a mixed deciduous Beech forest. Removed trees are marked with an asterisk in the tree list (B=Beech, A=Ash, M=Maple).

It is practically impossible for a forester to base his removal decisions on a comparison of all the trees with similar attributes occurring in the forest. The decision is only concerned with a particular cell involving a group of trees. The trees within each cell carry cell-specific selection preferences, and a thinning represents the sum of individual decisions which are spatially defined.

The simplest approach would involve a subdivision of the total forest area into a number of square cells containing a group of trees for which a collective decision can be made. This approach was followed by Daume et al. (1997) and is demonstrated in Fig. 5-24. The 30 x 30 m plot is subdivided into 4 square cells of equal size, each containing 8 trees on average. The attached list presents for each tree the number, species and preference values calculated according to Eq. 5-26. The data are grouped according to cells, and the trees in each cell are sorted with regard to their specific Pr_{ij}-value. The resulting prediction is very accurate.

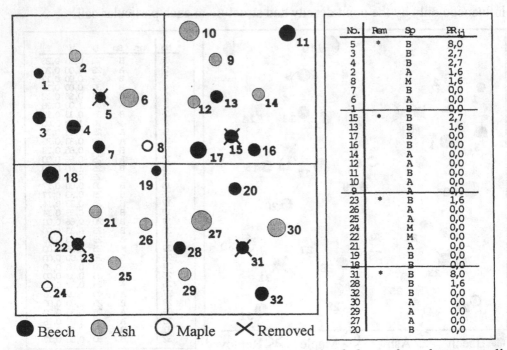

No.	Rem	Sp	PR$_{ij}$
5	*	B	8,0
3		B	2,7
4		B	2,7
2		A	1,6
8		M	1,6
7		B	0,0
6		A	0,0
1		B	0,0
15	*	B	2,7
13		B	1,6
17		B	0,0
16		B	0,0
14		A	0,0
12		A	0,0
11		B	0,0
10		A	0,0
9		A	0,0
23	*	B	1,6
26		A	0,0
25		A	0,0
24		M	0,0
22		M	0,0
21		A	0,0
19		B	0,0
18		B	0,0
31	*	B	8,0
28		B	1,6
32		B	0,0
30		A	0,0
29		A	0,0
27		A	0,0
20		B	0,0

● Beech ◍ Ash ○ Maple ✕ Removed

Figure 5-24. 30 x 30m section of the forest subdivided into four thinning cells (left); trees sorted according to the value of the preference index, separately for each thinning cell (right).

The example shows that the use of preference values based on two structure variables may be a very good basis for modelling the thinning process, provided that at least 3 neighbors are used. The identification of removed candidates may be further improved by using additional spatial information, such as preference values within thinning cells. The removed trees are those which carry the highest structural preference within their particular cells. This approach, based on the simplistic assumption that thinning cells are square, appears to produce good results, but further research is needed dealing with the optimum cell shapes and sizes, using data from a variety of forest types.

Prescription

The normative character of a thinning decision is a basic precondition for a realistic modelling approach. It assumes that the selection of trees for removal is based on consistent rules. This assumption is confirmed by empirical studies showing that a given thinning prescription may obviously be interpreted in a variety of ways, producing different results (Zucchini and Gadow, 1995; Füldner et al., 1996). However, consistent decision-making can be observed when a thinning is repeated by the same person (Kahle, 1995), an observation which quite naturally gives rise to the application of rule-based methods.

Few areas of science have raised such high expectations and at the same time have met with so much sceptical resistance as the area of artificial intelligence and its most important derivative, the rule-based system. Rules of thumb, based on practical experience, have always played an important part in the practice of forest management, but it is only recently that they have become acceptable as a subject of scientific investigation.

Many phenomena can be adequately described by a relation between objects, commonly known as a mathematical model. For example, the following statement relates the *relative spacing* (RS), a measure of density in even-aged stands, to the objects N and H:

$$RS = \frac{\sqrt{10\,000/N}}{H}$$

5-27

with   N = stems per ha;
                    H = dominant tree height [m].

The stems per ha corresponding to a given stand density is thus defined by:

$$N = \frac{10\,000}{(RS \cdot H)^2} \qquad \qquad 5\text{-}28$$

An *equation* is a quantitative *statement*, representing a relation between objects. An *implication*, commonly known as a rule, is a *relation between statements* and thus a higher-order concept. Consider the following rule which specifies that if the density exceeds an RS-value of 0.20 and the thinning is not risky, then the number of trees to be removed must be such that RS after the thinning will be equal to 0.3:

$$\left(\begin{array}{c} \dfrac{\sqrt{10\,000/N}}{H} < 0.20 \\ \& \\ \text{thinning is not risky} \end{array} \right) \Rightarrow \left[\begin{array}{c} \text{remove} \left\{ N - \dfrac{10000}{(0.30 \cdot H)^2} \right\} \text{trees / ha} \\ \text{in a thinning from below} \end{array} \right] \qquad 5\text{-}29$$

with   N = stems per ha;
              H = dominant tree height [m].

The world of forest management and silviculture is full of rules, many of which prescribe some ideal state which should be attained and then maintained, or some ideal sequence of operations which should be followed. Examples of ideal states are balanced age class distributions and stable forest structures, examples of idealized developments are prescribed silvicultural programs.

Ideal states and developments cannot always be attained, and therefore, the most effective rules are those which recognize that the conditions in the real world are usually suboptimal. Realistic rules which are oriented towards the actual forest state have a greater chance of being followed than those which only present an idealistic state. An example of a reality-based system of thinning rules is the concept presented in Tab. 5-10.

	H				
Final stem number/ha	**35m**	300	350	400	500 - 600
		Rest Period		*no*	*Fellings*
	27m	*Low Thinning*		*Low Thinning*	*Low Thinning*
Z-trees H/D-value	**21m**	300 - 350 70 - 80	250 - 300 80 - 85	200 - 250 85 - 90	(< 100?) (> 90 ?)
	15m	*Selective Thinning from above*			Uneconomic Thinning done
	10m	*Selective Thinning from above*		Uneconomic Thinning done	Uneconomic Thinning omitted again
		Selective Thinning	done	very late	
Uneconomic Thinning	**5m**	not necessary	carried out timely	omitted	omitted	omitted
Stems per ha Planted	**0m**	Normal	Too High	Too High	Too High	Too High

Table 5-10. System of thinning rules for even-aged Spruce stands in Austria based on dominant stand height (H) according to Johann and Pollanschütz (1981). Z-trees are the elite trees, the best individuals in a stand, characterized by exceptional vigor and quality; HD is the diameter/height ratio of a tree. Shading indicates increasing risk of instability.

Tab. 5-10 can be used very effectively for deriving a specific development path for a stand with a given treatment history. Consider, for example, a Spruce stand with a dominant height of 5m. The number of stems at planting was too high, but a timely uneconomical stem number reduction was done before reaching a dominant height of 5m. The future development path thus specifies a series of heavy selective thinnings until reaching a dominant height of about 20m. The objective is to provide sufficient growing space for the 300-350 Z-trees, the elite trees characterized by exceptional vigor and good quality. The stability of the Z-trees is very high, with a diameter/height ratio[1] of 70-80. The heavy high

---

[1] The diameter/height ratio *HD* of a tree represents the tree height [m] as a percentage of the breast height diameter [cm].

thinnings are followed by a series of moderate low thinnings and a rest period. The final crop of 300 trees per ha, consisting entirely of valuable Z-trees, are harvested when a dominant height of 35m is reached. Similarly, the future development path for a stand with a dominant height of 20m and too many trees per ha will consist of a few moderate low thinnings. The final crop contains less than 100 Z-trees and a high number of relatively small trees.

Silvicultural programs based on the Z-tree concept are based on successive stem number reductions with the aim of removing the Z-tree competitors. The number of trees removed in the vicinity of a Z-tree depends on the size of the elite tree and the sizes and distances of the competitors. Threshold distances to neighboring trees can be calculated when the tree coordinates are known. The following critical distance was proposed by Johann (1982):

$$G_{iz} = \frac{h_z}{A} \cdot \frac{d_i}{d_z} \qquad\qquad 5\text{-}30$$

with G_{iz} = critical distance between a Z-tree and a competitor,
 h_z = height of the Z-tree,
 d_i, d_z = breast height diameter of competitor i and Z-tree,
 A = parameter which defines the thinning weight.

A potential competitor is removed if its distance to the Z-tree E_{iz} is less than the critical distance G_{iz}. This simple rule can be implemented, not only in a computer program, but also in the forest. The crital distance to a Z-tree with a given height and diameter depends on the value of the thinning parameter A. The thinning weight decreases with decreasing critical distance, i.e. with increasing value of A.

The method may be illustrated using the hypothetical constellation in Fig. 5-25. The diagram shows a Z-tree with a height of 30m and a breast height diameter of 40 cm. Neighbor No. 1 with a breast height diameter of 40cm is located at a distance of 5m, the second neighbor is only 4m away with a diameter of 20cm. The value of the A-parameter has been set at 5.

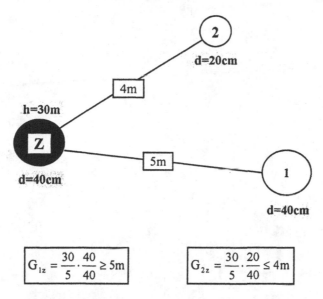

$$G_{1z} = \frac{30}{5} \cdot \frac{40}{40} \geq 5m \qquad G_{2z} = \frac{30}{5} \cdot \frac{20}{40} \leq 4m$$

Figure 5-25. Hypothetical constellation involving a Z-tree and two neighbors with distances and DBH's. The critical distances are shown below, involving an A-parameter value of 5.

The critical distance to the first neighbor is equal to 30/5 = 6, which is greater than the observed distance of 5m. Neighbor 1 is located within the critical area and will be removed. The critical distance to the second neighbor is only 3m, the actual distance 4m. Tree No. 2 is located outside the critical area and remains. Both trees would be removed if the A-value was smaller than 3.75, both would remain if the value was greater than 6.

Chapter Six

Model evaluation

Growth models are used for a variety of purposes. They are required to forecast
the development of a forest comprising various stands with different species, site
attributes and silvicultural regimes. They may be used for updating information
stored in a data base and for generating data that enable foresters to evaluate
alternative silvicultural regimes. Unreliable predictions are at the very least a
cause for concern, affecting supply forecasts and profitability calculations.
Evaluation is therefore a very important part of growth modelling. The different
approaches for evaluating model predictions include qualitative and quantitative
examinations (Soares et al., 1995).

Qualitative Evaluations

The main objective of the qualitative evaluation of a growth model is to examine
whether the performance of the individual components and of the model as a
whole is logically consistent and in agreement with current understanding of
elementary biological processes and the expected response of a forest to various
silvicultural treatments.

An illustrative example of a qualitative evaluation of an individual tree basal area model for Longleaf Pine (*Pinus palustris*) is presented by Quicke et al. (1994). Basal area growth of a tree (Δg) is predicted using the following general equation:

$$\Delta g = (BA\ submodel) \cdot (BAL\ multiplier) \cdot (d - Age\ multiplier)$$

The individual subcomponents were all fitted simultaneously to the data, thus minimizing overall model errors. The *G submodel* describes tree basal area increment as a function of stand basal area: $\Delta g = 11.52 \cdot e^{-0.0897 \cdot G^{0.5}}$. Assuming constant age, diameter and competitive position of the tree, an increase in individual tree basal area growth is predicted with decreasing stand basal area, as depicted in the left part of Fig. 6-1. Whereas the first model component describes the effect of overall density, the second is concerned with the competitive position of the tree within the population. This is expressed as the sum of the basal areas of all trees larger than the subject tree (BAL). The *BAL multiplier* is a function of BAL. Theoretically, the function $BAL\ multiplier = e^{-0.003974 \cdot BAL}$ may assume values between zero and one, as shown in the second diagram of Fig. 6-1.

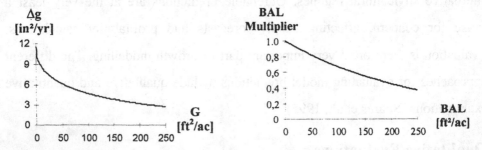

Figure 6-1. Two model components. Left: the effect of stand basal on tree basal area increment; Right: the effect of competitive position on the value of the BAL multiplier.

The basal area growth of a tree is reduced by increasing stand density and furthermore by an increasing sum of the basal areas of all trees larger than itself, assuming a constant tree diameter and age. This behaviour is consistent with existing empirical evidence.

The effects of tree diameter and tree age are described by the third model component, again an exponential term which can be decomposed into two elements: the Age multiplier $= e^{\beta_1 \cdot age}$ and the diameter term $\beta_1 = 0.2965 \cdot \{1 - e^{-0.3577 \cdot D}\} - 0.303$. The parameter β_1 determines the rate at which the age multiplier approaches zero (Fig. 6-2, diagram on the left). The third coefficient (-0.303) defines the y-axis intercept, the upper asymptote is determined by the difference between the first and the third coefficient (0.2965-0.303), while the second coefficient describes the rate at which this asymptote is reached.

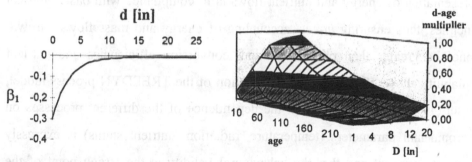

Figure 6-2. Effects of tree diameter d [inches] and age on the d-age multiplier.

As may be expected, the d-age multiplier decreases with increasing stand age and decreasing tree diameter, with G and BAL remaining at a constant level (Fig. 6-2, diagram on the right).

There is no predefined potential growth function imposed on the model, which is uncommon. The need to define a separate population for modelling potential growth is avoided by having a single equation with all parameters estimated simultaneously. This is a special feature, adding consistency and reducing overall error, but requiring a special evaluation involving a comparison of model performance with free-growing trees. The ability of the model to simulate reasonable growth for open-grown trees was tested by imposing an open grown condition with stand basal area equal to tree basal area and BAL equal to zero. A test involving a comparison of the simulated diameter development with the observed diameters of 81 open grown-trees showed that the model was capable of providing a good approximation of potential growth. The model logic specifies that maximum growth rates result from the interaction between stand basal area, stand age and tree diameter with the BAL multiplier equal to 1 for the largest tree in the stand.

The qualitative evaluation examines whether model behaviour outside the range of the available data is reasonable, whether signs and values of coefficients are plausible and, in the case of biological process models, whether the representation of energy and nutrient flows is in compliance with basic physical principles, thus ensuring *honest* accounting of energy and mass flows. Growth cannot be greater than photoproduction, conversion efficiencies and nutrient availability allow. Referring to the evaluation of the TREEDYN process model, Bossel (1994, p. 112) states that the dependence of the different processes on environmental parameters (temperature, radiation, nutrient status) is expressly taken into account and that the behavioural validity is the strong point of the model.

A thorough qualitative model evaluation is the appropriate measure to make sure that the various process formulations are neither too simple nor inadequate. Making the model more complex, however, requires more parameters, causing numerous additional interactions between individual processes. This is the paradox of modelling. Greater detail improves understanding and realistic description of real processes, but this is achieved at the cost of added model complexity. It is difficult to accept a model that cannot be evaluated. Thus, the appropriate level of complexity is defined by the ability to present a comprehensive model evaluation.

Quantitative Evaluations

A quantitative evaluation of a growth model may include a characterization of errors in terms of their magnitude and the distribution of residuals; tests for bias and precision; and a sensitivity analysis to identify those components which have the greatest influence on the predictions (Soares et al., 1995; Vanclay, 1994, p. 206 et sqq.).

The standard linear regression assumptions are that the random errors ε of the model $Y = \alpha + \beta X + \varepsilon$ are additive, independent and normally distributed with zero mean and unknown but constant variance. These assumptions represent an ideal which is not always attained.

Characterizing Model Error

Errors, characterized by their magnitude and the distribution of residuals, may be assessed by comparing a model with a given set of observations. Important concepts in model evaluation are the *bias* and the *precision*, which determine the *accuracy* of a prediction. These concepts, used by Freese (1960) in the context of forest sampling, are illustrated in Fig. 6-3.

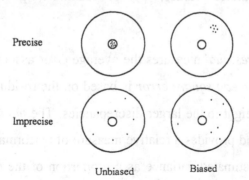

Figure 6-3. An imaginary target and the distribution of hits, illustrating the concepts of precision and bias. A precise and unbiased prediction is accurate.

The *mean residual MRES*, a measure of average model **bias**, describes the directional magnitude, i.e. the size of expected under- or overestimates. Indices of model **precision** are the *absolute mean residual* (or mean absolute discrepancy), the *root mean square error*, the *model efficiency* and the *variance ratio* (Tab. 6-1).

Criterion	Formula	Ideal Value
Mean residual	$MRES = \dfrac{\sum(y_i - \hat{y}_i)}{n}$	0
Absolute mean residual	$AMRES = \dfrac{\sum\lvert y_i - \hat{y}_i\rvert}{n}$	0
Root mean square error	$RMSE = \sqrt{\dfrac{\sum(y_i - \hat{y}_i)^2}{n-1-p}}$	0
Model efficiency	$MEF = \dfrac{\sum(y_i - \hat{y}_i)^2}{\sum(y_i - \bar{y})^2}$	0
Variance ratio	$VR = \dfrac{\sum(\hat{y}_i - \bar{\bar{y}})^2}{\sum(y_i - \bar{y})^2}$	1

Table 6-1. Four criteria for evaluating model performance. MRES measures bias, the remaining three criteria indicate model precision (y=observed values; ŷ =predicted values; (y-ŷ)=residuals; p=number of model parameters).

The *absolute mean residual* measures the average error associated with a single prediction. The *root mean square error* is based on the residual sum of squares, which gives more weight to the larger discrepancies. The *model efficiency* index is analogous to R^2 and provides a relative measure of performance. The *variance ratio* measures the estimated variance as a proportion of the observed one. The criteria *mean residual*, *absolute mean residual* and *root mean square error* may be expressed as relative values, which is more revealing when components with different measurement units are compared.

Obviously, the number of indices that can be devised to measure model performance is boundless. There are numerous possibilities to measure model precision, and it is very difficult to rank the different indices, as they measure different things. Basically, a good index is one which provides interpretable reference values, such as an interpretable maximum or minimum or an optimum as shown in Table 6-1.

One of the most common procedures for evaluating a model is to examine the residuals for all possible combinations of variables. The aim is to detect dependencies or patterns that indicate systematic discrepancies. Observed values may be plotted over predicted values, or residuals over observed values. Fig. 6-4 presents an example.

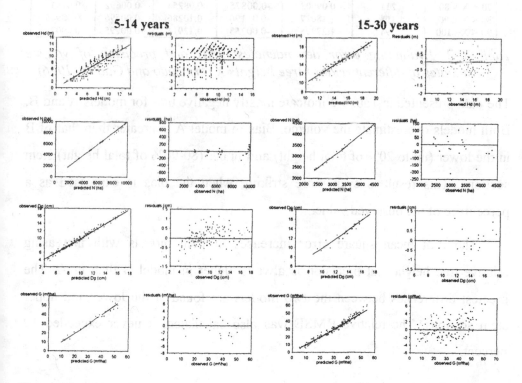

Figure 6.4. Visual analysis of residuals. Left: observed values (y) plotted over predicted values (ŷ). Right: residuals (y − ŷ) plotted over observed values (y).

An important aim of model evaluation is to detect error dependencies. Errors may be influenced by projection length, stand age or tree height. The criteria *mean residual* (for evaluating bias) and *root mean square error* (for evaluating precision) were used by Trincado and Gadow (1996) to evaluate the accuracy of the predictions of the cumulative volume of two taper functions at different relative tree heights (Table 6-2).

Relative Tree Height (%)	Number of Observations	Mean Volume (m³)	Model A		Model B	
			MRES(m³)	RMSE(m³)	MRES(m³)	RMSE(m³)
10 < X ≤ 20	191	0.30865	-0.01455	0.02231	-0.00243	0.00846
20 < X ≤ 30	203	0.45857	-0.00323	0.01912	0.00078	0.01498
30 < X ≤ 40	205	0.59990	0.00003	0.02954	0.00132	0.02614
40 < X ≤ 50	196	0.73685	0.00152	0.04167	-0.00094	0.04079
50 < X ≤ 60	198	0.87749	-0.00411	0.05664	-0.00430	0.05524
60 < X ≤ 70	203	0.99454	-0.00817	0.07059	-0.00912	0.06997
70 < X ≤ 80	211	0.09903	-0.00875	0.08256	-0.00692	0.07853
80 < X ≤ 90	198	1.18017	-0.01196	0.10286	-0.00596	0.08945
90 < X <100	194	1.27298	-0.00753	0.12914	-0.00295	0.09077

Table 6-2. Evaluating error dependencies: bias and precision of volume estimates at different relative tree heights X (Trincado and Gadow, 1996).

The data presented in Tab. 6-2 indicate mostly negative bias for models A and B. Both models overestimate the volume. Bias in model A is greater than that of B in the lower (up to 20% of total height) and upper (80-90% of total height) stem sections. This result is even more striking when the bias is expressed as a percentage of the observed value.

The root mean square error increases in both models with increasing relative tree height, but precision is always higher in model B than in A. The greatest differences between the two models are found at the lower and upper stem sections. The relative RMSE was also calculated, it never exceeded 10 percent.

Statistical Tests

Formal statistical tests are available for evaluating model performance. Linear regression of predicted on observed values is used to test for bias while a standardized sums of squares criterion may be used in a test for evaluating precision (Tab. 6-3).

Criterion	Formula	Ideal Value
Linear regression	$y = \alpha + \beta \cdot \hat{y}$	$\alpha=0; \beta=1$
Model precision	$MPR = \dfrac{1}{\sigma^2} \sum (y_i - \hat{y}_i)^2$	0

Table 6-3. Two criteria used in formal statistical tests. Linear regression of predicted on observed values is used to test for bias. The MPR criterion proposed by Freese (1960) may be used in a test for evaluating precision (σ^2 is the variance of y).

A useful test for bias is the simultaneous F-test for slope and zero intercept, as applied by Hui and Gadow (1993b)[1]. The test is based on a linear regression of observations (y_i) against predictions ($\hat{y}_i$) of the form $y = \alpha + \beta \cdot \hat{y}$. For testing the hypothesis that $\alpha=0$ and $\beta=1$, the test criterion F is calculated as follows:

$$F_{2,n} = \frac{(n-2)\left[\left(a\sum y_i + b\sum \hat{y}_i y_i\right) - \left(2 \cdot \sum \hat{y}_i y_i - \sum \hat{y}_i^2\right)\right]}{2 \cdot \left[\sum y_i^2 - \left(a\sum y_i + b\sum \hat{y}_i y_i\right)\right]} \qquad 6\text{-}1$$

where n = number of sampling units.

F_α is the threshold value of F (normally at $\alpha=5\%$ or 1%). The hypothesis is rejected, when $F_{2,n} > F_\alpha$.

---

[1] See also Dent and Blackie (1979); Liang and Tang (1989); Mayer and Butler (1993).

Symbols used

μ	distribution mean
σ^2	distribution variance
A_i	influence zone of subject tree i
a_{ij}	influence zone overlap between subject tree i and competitor j
B°	degree of stocking (basal area per hectare expressed as a proportion of the normal basal area, defined by a yield table)
CA	crown surface area [m²]
CI	competition index as defined by Bella
ΔCI	change of CI caused by thinning
CL	crown length [m]
CR	crown ratio [crown length/tree height]
CV	crown volume [m³]
d	individual tree breast height diameter [cm]
Dg	quadratic mean diameter of a stand of trees [cm]
$\overline{D}$	average stand diameter [cm]
d_{1i}, d_{2i}	breast height diameter of the i-th tree [cm] at ages t_1 and t_2
$dist_{ij}$	distance between a reference tree i and its neighbor j [m]
d_h	stem diameter [cm] at a given tree height h
dkra	diameter at base of crown
d_m	minimum Dg yielding products with a thin-end diameter m [cm]
dmax	maximum crown diameter
G	basal area [m²/ha]
ΔG	annual basal area growth [m²/ha]
Δg_i	future 5-year basal area growth of a tree [cm²]
G_1, G_2	basal area [m²/ha] corresponding to stand ages t_1 and t_2
$G_{thn,}$	basal area per ha removed during a thinning operation
G_{tot}	basal area per ha before the thinning operation

H	dominant stand height [m]
$\overline{\overline{H}}$	average stand height [m]
H_1, H_2	dominant stand height at ages t_1 and t_2
H_{100}	dominant stand height at age 100 [m]
h	individual tree total height [m]
Δh	height growth [m]
$Δh_{pot}$	potential height growth [m]
HLC	height to base of crown [m]
KA	height to crown base [m]
KR	crown radius [m]
KM	crown surface area [m²]
l	survival rate of a tree with breast height diameter d
L	total survival rate in a population of trees
Lo	length of crown exposed to sunlight [m]
M_i	*Mingling*, the relative proportion of neighboring trees of tree i, which differ from i with regard to species
m	thin-end diameter of a stem section [cm]
Nmax	maximum number of surviving trees per ha
N_1, N_2	stems per ha corresponding to stand ages t_1 and t_2
N_{thn},	number of stems per ha removed during a thinning operation
N_{tot}	number of stems per ha before the thinning operation
OVS	overstocking multiplier
removed p_{ij}	proportion of trees removed in structure class ij
before p_{ij}	proportion of trees in structure class ij, before the thinning
PR_{ij}	thinning selection preference in structure class ij [i=1..m, j=1..n]
PrCI	competition index based on *shading* and *constriction*
R_d	volume ratio to a given thin-end diameter m [cm]
r(E)	crown radius [m] at distance E m from the tip

tree radius at height h' [cm]

reference height [m]

crown radius in the upper and shaded parts respectively [m]

maximum crown radius [m]

radius at crown base [m]

segregation parameter

distance between two neighboring trees [m]

site index (dominant stand height at reference age t years)

location separation parameter

scale separation parameter

SG-ratio: $SG = \dfrac{(N_{thn}/N_{tot})}{(G_{thn}/G_{tot})}$

size-weighting factor (S = tree height multiplied by the potential crown radius of a solitary tree)

site quality class

differentiation of breast height diameters for a given tree i (i=1...I) and its n nearest neighbors j (j=1..n)

stand age [years]

age at which a tree reaches breast height (1,3 m)

volume of growing stock [m³/ha]

remaining stand volume at age 100 [m³/ha]

volume of the remaining stand [m³/ha]

volume of logs with a minimum thin-end diameter *m* [m³/ha]

total stand volume [m³/ha]

individual tree volume [m³]

predictor variable, e.g. stand age

some measure of tree dimension or stand production

volume [m³] over bark to a given thin-end diameter m [cm] and with a given diameter at the base of the crown *dkra* [cm]

over bark volume from the base of the tree to the base of the crown [m³]

ΔW estimated change of size variable W

ΔW_{pot} potential change of size variable W

ε_i index of constriction calculated for tree i

$\omega_i^{Sp}, \omega_i^{Be}$ index of shading caused by neighboring spruce (*Picea abies*) or
                          beech (*Fagus sylvatica*) trees, respectively, for tree i

Literature

Abetz, A., 1975: Eine Entscheidungshilfe für die Durchforstung von Fichtenbeständen. AFZ 30 (33/34): 666-667.

Aguirre, O., 1989: Aufstellung von Ertragstafeln auf der Basis einmaliger Waldaufnahmen am Beispiel von *Pinus pseudostrobus* im Nordosten Mexikos. Dissertation. Uni Göttingen, Forstwiss. Fachbereich.

Akça, A., 1993: Zur Methodik und Bedeutung der kontinuierlichen Forstinventuren. AFJZ 11: 193-198.

Akça, A., Gadow, K. v., Mench, A., Mann, P., Pahl, A. and Setje-Eilers, U., 1994: Überprüfung des Formquotienten q_7 (d_7/BHD) in der Bundeswaldinventur für die Hauptbaumarten Buche, Eiche, Fichte und Kiefer im landeseigenen Wald Nordrhein-Westfalens. Project report: 23 p.

Albert, M., 1997: Positionsabhängige Einzelbaummodellierung am Beispiel des Simulationsansatzes von Hasenauer. Manuscript, Institut. f. Forsteinr. u. Ertragsk. d. Univ. Göttingen: 6 p.

Alder, D., 1979: A distance-independent tree model for exotic conifer plantations in East Africa. For. Sci., 25(1): 59-71.

Alder, D., 1980: Estimación del volumen forestal y predicción del rendimiento. Estudios FAO. Montes 22/2. Roma.

Álvarez González, J.-G., 1997: Analisis y caracterización de las distribuciones diámetros de *Pinus pinaster* Ait Galicia. Diss. Univ. Madrid.

Amaro, A., Tomé, M. and Reed, D., 1997: Modelling dominant height growth - Eucalyptus plantations in Portugal. Unpubl. manuscript: 15 p.

Amateis R.L., Burkhart, H.E. and Burk, T.E., 1986. A ratio approach to predicting merchantable yields of unthinned loblolly pine plantations. For. Sci. 32: 187 - 296.

Ammon, W., 1951: Das Plenterprinzip in der Waldwirtschaft. Paul Haupt, Bern-Stuttgart.

Anderson, M.C., 1964: Studies of the woodland light climate. Journal of Ecology. 52: 27-41.

Arney, J.D., 1973: Tables for quantifying competitive stress on individual trees. Can. For. Serv. Inf. Report BC-X-78.

Arnswaldt, H.J. v., 1950: Wertkontrolle in Laubholzrevieren. Forstarchiv: 130-135.

Arnswaldt, H.J.v., 1953: Wertkontrolle. AFZ 8, S.408-410.

Assmann, E. and Franz, F., 1965: Vorläufige Fichten-Ertragstafel für Bayern. Forstw. Centralblatt 84: 13-43.

Assmann, E., 1953: Zur Bonitierung süddeutscher Fichtenbestände. AFZ 10: 61-64.

Bachmann, P., 1990: Produktionssteigerung im Wald durch vermehrte Berücksichtigung des Wertzuwachses. Eidgenössische Forschungsanstalt für Wald, Schnee und Landschaft, Birmensdorf.

Bailey, R.L. and Dell, T.R., 1973: Quantifying diameter distributions with the Weibull function. For. Sci. 19: 97-104.

Baur, F., 1877: Die Fichte in Bezug auf Ertrag, Zuwachs und Form. Berlin.

Bella, I.E., 1971: A new competition model for individual trees. For. Sci. 17: 364-372.

Bernstorff, A. Graf v. and Kurth, H., 1988: Betriebswerk Gartow. Working plan, Gartow forest.

Bertalanffy, L. von, 1948: Quantitative laws in metabolism and growth. Quart. Rev. Biol. 32: 217-230.

Biber, P., 1996: Konstruktion eines einzelbaumorientierten Wachstumssimulators für Fichten-Buchen-Mischbestände im Solling. Diss. Forstw. Fak. d. Ludwig-Maximilians-Univ. München: 239 p.

Biging, G. and Wensel, L.C., 1985: Site index equations for young-growth mixed conifers of Northern California. N. California Forest Yield Coop. Res. Note 8. Univ. of California, Berkeley.

Biging, G.S. and Dobbertin, M., 1992: A comparison of distance-dependent competition measures for height and basal area growth of individual conifer trees. For. Sci. 38 (3): 695-720.

Biging, G.S. and Dobbertin, M., 1995: Evaluation of competition indices in individual tree growth models. Forest Science 41(2): 360-337.

Boardman, R. 1984: Fast-growing species-pattern, process and ageing. Proceedings: IUFRO Symposium on site and productivity of fast growing plantations, Pretoria:49.

Borggreve, B., 1891: Die Holzzucht. 2nd edition, Berlin.

Bossel, H., 1994: TREEDYN3 forest simulation model. Report, Forest Ecosystems Research Centre, Göttingen, series B, vol 35: 118p.

Botkin, D.B., -1993: Forest dynamics - an ecological model.. Oxford University Press.

Brandl, H., 1992: Die Entwicklung der Durchforstung in der deutschen Forstwirtschaft. AFJZ 163 (4): 61-70.

Brink, C. and Gadow, K.v., 1986: On the use of growth and decay functions for modelling stem profiles. EDV in Medizin u. Biologie 17(1/2): 20-27.

Burk, T.E. and Newberry, 1984: A simple algorithm for moment-based recovery of Weibull distribution parameters. For. Sci. 30 (2): 329-332.

Burger, H., 1939: Kronenaufbau gleichaltriger Nadelholzbestände. Mitt. Schweiz. Anst. f. d. forstl. Vers.Wesen 21. S.5-58.

Burkhart, H. E., 1977: Cubic-foot volume of loblolly pine to any merchantable top limit. S. J. Appl. For. *1(2)*, 7-9.

Burkhart, H. E., 1987: Data collection and modelling approaches for forest growth and yield prediction. In: Predicting Forest Growth and Yield - Current Issues, Future Prospects. Inst. of Forest Resources. Univ. of Washington. Contribution Nr. 58: 3-16.

Cao, Q. V. and Burkhart, H. E., 1984: A segmented distribution approach to modeling diameter frequency data. For. Sci. 30: 129-137.

Clark, P.J., Evans, F.C., 1954: Distance to nearest neighbor as a measure of spatial relationships in populations. Ecology, 35: 445-453.

Clutter, J. L. and Lenhart, J.D., 1968: Site index curves for old-field loblolly pine plantations in the Georgia Piedmont. Georgia Forestry Res. Council, Rep. 22..

Clutter, J. L., 1963: Compatible growth and yield models for loblolly pine. For. Sci., 9:354-371.

Clutter, J.L. and Allison, B.J., 1974: A growth and yield model for Pinus radiata in New Zealand. In: J. Fries (ed): Growth models for tree and stand simulation. Proc. IUFRO meeting S4.01.04, Stockholm.

Clutter, J.L., 1980: Development of taper functions from variable-top merchantable volume equations. For. Sci. 26(1): 117-120.

Clutter, J.L. and Jones, E.P., 1980: Prediction of growth after thinning in old field slash pine plantations. USDA For. Serv. Res. Paper SE-217.

Clutter, J.L., Fortson, J.C., Pienaar, L.V., Brister, G.H. and Bailey, R.L., 1983. Timber management - a quantitative approach. John Wiley.

Courbaud, B, 1995: Modélisation de la croissance en forêt irrégulière, - perspectives pour les pessières irrégulières de montagne. Rev. For. Fr. XLVII No. sp.: 173.

Craib, I.J., 1939: Thinning, pruning and management studies on the main exotic conifers grown in South Africa. Govt. Printer, Pretoria.

Dagnelle, P., Palm, R., Rondeux, J. and Thill, A., 1988: Tables de production relatives a l'epicea commun. Les Presses Agronomiques de Gembloux, Belgium: 122 p.

Daniels, R.F., 1976: Simple competition indices and their correlation with annual loblolly pine tree growth. Forest Science 22(4): 454-456.

Daume, S., 1995: Durchforstungssimulation in einem Buchen-Edellaubholz-Mischbestand. Diplomarbeit, Forstliche Fakultät, Universität Göttingen.

Daume, S., Füldner, K. and Gadow, K.v., 1997: Zur Modellierung personenspezifischer Durchforstungen in ungleichaltrigen Mischbeständen (*AFJZ*, in press).

Degenhardt, A., 1995: Analyse der Entwicklung von Bestandesstrukturen mit Hilfe des Modells der zufälligen Punktprozesse in der Ebene. Tagungsbericht, Dt. Verb. Forstl. Forsch. Anst., Sektion Biometrie u. Informatik, 8.-10.9.93 in Freising: 93-105.

Dengler, A., 1982: Waldbau auf ökologischer Grundlage. Edited by E. Röhrig und H.A. Gussone. Paul Parey, Hamburg and Berlin.

Dent, J.B. and Blackie, M.J., 1979: Systems simulation in agriculture. Applied Science Publishers, London.

Demaerschalk, J.P., 1973: Integrated systems for the estimation of tree taper and volume. Can. J. For. Res. 3(1): 90-94.

Dobson, A.J., 1990: An introduction to generalized linear models. Capman and Hall. London, New York.

Donnelly, K.P., 1978: Simulations to determine the variance and edge effect of total nearest-neighbour distance. In: Simulation Methods in Archaeology. Cambridge University Press, London: 91-95.

Dougherty, E.R., 1990. Probability and statistics for the engineering, computing and physical sciences. Prentice-Hall.

Dralle, K., 1997: Locating trees by digital image processing of aerial photos. Dina Research Rapport No. 58: 116 p.

Düser, R., 1978: Programmierte Berechnung stehenden Holzes mit dem Taschenrechner nach der Düser-Flori-Formel. Beilage der AFZ, Nr. 36.

Ek, A.R. and Monserud, R.A., 1974: FOREST - a computer model for simulating the growth and reproduction of mixed-species forest stands. Univ. Wisc. School Nat. Resources Res. Rep. R2635.

Ford, E.D. and Diggle, P.J., 1981: Competition for light in a plant monoculture modelled as a spatial stochastic process. Annals of Botany 48: 481-500.

Forss, E., Gadow, K. v. and Saborowski, J., 1996: Growth models for unthinned *Acacia mangium* plantations in South Kalimantan, Indonesia. J. of Trop. For. Sci. 8 (4): 449-462.

Franz, F., 1965: Ermittlung von Schätzwerten der natürlichen Grundfläche mit Hilfe ertragskundlicher Bestimmungsgrößen des verbleibenden Bestandes, Forstw. Cbl. 84: 357-386.

Franz, F., 1972: Ertragskundliche Prognosemodelle. Fw.Cbl. 91: S.65-80.

Freese, F., 1960: Testing accuracy. For. Sci. 6: 139-145.

Füldner, K. and Gadow, K. v., 1994: How to define a thinning in a mixed deciduous beech forest. Proc. IUFRO Conference in Lousa, Portugal: Mixed stands - research plots, measurements and results, models, 1994: 31-42.

Füldner, K., 1995: Strukturbeschreibung von Buchen-Edellaubholz-Mischwäldern. Dissertation, Forstliche Fakultät, Göttingen. Cuvillier Verlag, Göttingen.

Füldner, K., Sattler, S., Zucchini, W. and Gadow, K. v., 1996: Modellierung personenabhängiger Auswahlwahrscheinlichkeiten bei der Durchforstung. AFJZ 167: 159-162.

Gadow, K. v., 1984: The relationship between diameter and diameter increment in *Pinus patula*. Proceedings of the IUFRO Conference "Site and Productivity of Fast-growing Plantations", held at Pretoria 1984, Vol. 2: 741-751.

Gadow, K. v., 1987: Untersuchungen zur Konstruktion von Wuchsmodellen für schnellwüchsige Plantagenbaumarten. Schriftenreihe d. Forstw. Fak., Univ. München, No. 77: 147 p.

Gadow, K. v., 1992: Wachstums- und Ertragsmodelle für die Forsteinrichtung. Proc. annual meeting of the *Deutscher Verband Forstlicher Forschungsanstalten, Sektion Ertragskunde*, in Grillenburg.

Gadow, K. v. and Bredenkamp, B.V., 1992: Forest management. Academica, Pretoria.

Gadow, K. v. and Hui, G.Y., 1993: Stammzahlentwicklung und potentielle Bestandesdichte bei *Cunninghamia lanceolata*. Centralblatt für das gesamte Forstwesen 110 (2): 41-48.

Gadow, K.v., Füldner, K., 1995: Zur Beschreibung forstlicher Eingriffe. Forstw. Cbl. 114, S. 151-159.

Gaffrey, D., 1996: Sortenorientiertes Bestandeswachstums- Simulationsmodell auf der Basis intraspezifischen, konkurrenzbedingten Einzelbaumwachstums- insbesondere hinsichtlich des Durchmessers- am Beispiel der Douglasie. Dissertation, Forstl. Fak. Göttingen.

Gaffrey, D., 1988: Forstamts- und bestandesindividuelles Sortimentierungsprogramm als Mittel zur Planung, Aushaltung und Simulation. Diplomarbeit, Fachbereich Forstwissenschaft der Georg-August-Univ. Göttingen.

Gál, J. and Bella, I. E., 1994: New stem taper functions for 12 Saskatschewan timber species. Nat. Resour. Can., Can. For. Serv., Northwest Reg. North. For. Cent. Edmonton, Alberta. Inf. Rep. NOR-X-338, 25.

Garcia, O., 1984: New class of growth models for even-aged stands: *Pinus radiata* in Golden Downs Forest. N.Z. J. of For. Sci. 14: 65-88.

Garcia, O., 1988: Experience with an advanced growth modelling methodology. In: Ek, A.R., Shifley, S.R. & Burke, T.E. (eds): Forest growth modelling and prediction. USDA For. Serv. Gen. Techn. Rep. NC-120: 668-675.

Garcia, O., 1994: The state space approach in growth modelling. Can. J. For. Res. 24: 1894-1903.

Garcia, O., 1997: Another look at growth equations. Working paper, Royal Veterinary and Agricultural University, Copenhagen: 7 p.

Gerrard, D.J., 1969: Competition quotient - a new measure of the competition affecting individual forest trees. Mich. State Univ. Agr. Exp. Stn. Res. Bull. No. 20.

Goebel, V., 1992: Standortorientiertes Sortenmodell für die Baumart Buche im Forstamt Reinhausen. Diplomarbeit. Forstl. Fak. Göttingen.

Green, P.J. and Sibson, R., 1977: Computing dirichlet tesselations in the plane. The Computer Journal: 168-173.

Goulding, C. J., 1972: Simulation technique for a stochastic model of growth of Douglas-fir. Ph. D. thesis, Univ. of Brit. Col., Vancouver, 185 p.

Grundner, F. and Schwappach, A., 1942: Massentafeln zur Bestimmung des Holzgehaltes stehender Waldbäume und Waldbestände. 9. Auflage, Berlin.

Grut, M., 1970: *Pinus radiata* - growth and economics. Balkema, Cape Town.

Hafley, W.L. and Schreuder, H.T., 1977. Statistical distributions for fitting diameter and height data in even-aged stands. Canadian Journal of Forest Research 7, 481-487.

Hamilton, G.J. and Christie, J.M., 1971: Forest management tables (metrics). For. Comm. booklet n° 34. Londen: 32 p.

Hart, H.M.J., 1928: Stamtal en Dunning - een Orienteerend Onderzoek naar de Beste Plantwijdte en Dunningswijze voor den Djati. H. Venman & Zonen, Wageningen.

Hartig, R., 1868: Die Rentabilität der Fichtennutzholz- und Buchenbrennholzwirtschaft im Harze und im Wesergebirge. Stuttgart.

Hasenauer, H., Moser, M. and Eckmüllner, O., 1995: MOSES - a computer simulation program for modelling stand response. In: Pinto da Costa, M.E. and T. Preuhsler (eds.): Mixed stands, research plots, and results, models. Inst. Sup. De Agronomia, Univ. Tecnica de Lisboa, Portugal.

Heck, C.H., 1904: Freie Durchforstung. Berlin.

Hegyi, F., 1974: A simulation model for managing jackpine stands. In: Growth Models for Tree and Stand Simulation, Fries, J. (ed.). Royal Coll. of Forestry, Stockholm: 74-90.

Hengst, E., 1959: Die Schaftform der Weymouthskiefer. Wiss. Z. d. T.H. Dresden, Heft 4.

Holmes, M.J. and Reed, D.D., 1991: Competition indices for mixed species in Northern Hardwoods. Forest Science 37(5): 1338-1349.

Honer, T. G., 1965: A new total cubic foot volumen function. For. Chron. *41*, 476-493.

Hui, G. Y.and Gadow, K. v., 1993a: Zur Modellierung der Bestandesgrundflächenentwicklung dargestellt am Beispiel der Baumart *Cunninghamia lanceolata*. Allg.Forst- u. J.-Ztg., 164(8):144-145.

Hui, G. Y.andGadow, K. v., 1993b: Zur Entwicklung von Einheitshöhenkurven am Beispiel der Baumart *Cunninghamia lanceolata*. Allg.Forst- u. J.-Ztg., 164(12):218-220.

Hui, G. Y., and Gadow, K. v., 1996: Ein neuer Ansatz zur Modellierung von Durchmesserverteilungen. Cbl. F. d. ges. Forstwes. 113 (¾): 101-113.

Hui, G. Y., and Gadow, K. v., 1997: Entwicklung und Erprobung eines Einheitsschaftmodells für die Baumart *Cunninghamia lanceolata*. Forstw. Cbl. 116 (5): 315-321.

Jansen, J.J., Sevenster, J., and Faber, P.J., 1996: Opbrengsttabellen voor belangrijke boomsoorten in Nederland. Hinkeloord Report Nr. 17, Landbouwuniversiteit, Wageningen: 202 p.

Johnson, N.L. 1949a: Systems of frequency curves generated by methods of translation. Biometrika 36, 149-176.

Johnson, N.L. 1949b: Bivariate distributions based on simple translation systems. Biometrika 36, 297-304.

Johann, K. and Pollanschütz, J., 1981: Standraumregulierung bei Fichte und Buche, Mischwuchsregulierung. Seminarreihe d. Forstl. Bundesversuchsanstalt, Vienna: 51 pp.

Johann, K., 1982: Der A-Wert ein objektiver Parameter zur Bestimmung der Freistellungsstärke von Zentralbäumen. Tagungsbericht, Deutscher Verband Forstlicher Versuchsanstalten - Sektion Ertragskunde, Weibersbrunn. 146-158, 1982.

Kahle, M., 1995: Die Analyse von Bestandeseingriffen in einem Buchen-Edellaubholz-Mischbestand. Diplomarbeit, Forstliche Fakultät Göttingen.

Kahn, M., 1994: Modellierung der Höhenentwicklung ausgewählter Baumarten in Abhängigkeit vom Standort. Forstliche Forschungsberichte München, Nr. 141.

Kassier, H.W., 1993: Dynamics of diameter and height distributions in commercial timber plantations. PhD dissertation, Faculty of Forestry, Univ. of Stellenbosch, South Africa.

Kennel, R., 1972: Die Buchendurchforstungsversuche in Bayern von 1870 bis 1970. Forstliche Forschungsberichte München, No. 7: 264 p.

Kiviste, A.K., 1988: Growth functions for forests. Agricultural Academy, Estonia.

Knoebel, B.R. and Burkhart, H.E., 1991: A bivariate distribution approach to modeling forest diameter distributions at two points of time. Biometrics 47, 241-253.

Kolström, T., 1992: Dynamics of uneven-aged stands of Norway Spruce - a model approach. Dissertation. Joensun. Finnland.

Kozak, A., 1988: A variable exponent taper equation. Can. J. For. Res. 18: 1363-1368.

Kraft, G., 1884: Beiträge zur Lehre von den Durchforstungen, Schlagstellungen und Lichtungshieben. Hannover.

Kramer, H. and Akça, A., 1987: Leitfaden zur Waldmeßlehre. J.D. Sauerländer's Verlag, Frankfurt a.M.

Kramer, H., 1988: Waldwachstumslehre. Verlag Paul Parey, Hamburg.

Kramer, H., 1990: Nutzungsplanung in der Forsteinrichtung. 2nd ed., J. D. Sauerländer's Verlag,. Frankfurt/Main.

Laasasenaho, J., 1982: Taper curve and volume functions for pine, spruce and birch. Comm. Instituti Forestalis Fenniae No. 108, Helsinki.

Leary, R.A., 1979: Design. In: A generalized forest growth projection system applied to the Lake States region. USDA For. Serv. Gen. Tech. Rep. NC-49: 5-15.

Lee, D.T.. 1980: Two-dimensional Voronoi diagram in the LP-metric. Journal ACM: 604-618.

Lee, W. -K., 1993: Wachstums- und Ertragsmodelle für *Pinus densiflora* in der Kangwon-Provinz, Korea. Dissertation. Univ. Göttingen, Forstwiss. Fachbereich.

Lee, W.K. and Gadow, K. v., 1997: Iterative Bestimmung der Konkurrenzbäume in *Pinus densiflora* Beständen. AFJZ 168 (3/4): 41-44.

Leibundgut, H., 1978: Die Waldpflege. Paul Haupt, Bern and Stuttgart.

LeMay, V., Kozak, A., Muhairwe, K. and Kozak, R.A., 1993: Factors affecting the performance of Kozak's (1988) variable-exponent taper function. In: Wood, G.B. and. Wiant, H.V. (eds), 1993: Proc. IUFRO Conf. Modern methods of estimating tree and log volume, West Virginia University: 34-53.

Lemm, R., 1991: Ein dynamisches Forstbetriebssimulationsmodell. Diss. Professur für Forsteinrichtung und Waldwachstum der ETH Zürich.

Leslie, P.H., 1945: On the use of matrices in certain population mathematics. Biometrika 33: 183-212.

Lewandowski, A. and Pommerening, A., 1996: Waldsim - ein Programm zur Stichprobensimulation in strukturreichen Wäldern. Institut Forsteinrichtung, Georg-August-Univ. Göttingen: 31p.

Lewandowski, A. and Gadow, K.v.; 1997: Ein heuristischer Ansatz zur Reproduktion von Waldbeständen. AFJZ (in press)

Lewis, N.B., Keeves, A. and Leech, J.W. 1976: Yield regulation in South Australian *Pinus radiata* plantations. Woods and Forests Dept., South Australia, Bull. 23.

Liang, K. J. and Tang, Sh. Zh., 1989: IBM PC Program package, Forestry Publishers, Beijing, China.

Lorimer, C.G., 1983: Test of age-independent competition indices for individual trees in natural hardwood stands, For. Ecol. Management 6:343-360.

MacKinney, A.L., Schumacher, F.X. and Chaiken, L.E., 1937: Construction of yield tables for non-normal loblolly pine stands. J. Agric. Res. 54: 531-545.

Madrigal, A., Puertas, F. and Martinez Millan, F.J., 1992: Tablas de producción para *Fagus sylvatica* en Navarra. Gobierno de Navarra. Dpto. de Agric. Ganadería y Montes. Serie Agraria n° 3, Pamplona: 133 p.

Maltamo, M., Puumalainen, J. and Päivinen, R., 1995: Comparison of Beta and Weibull functions for modelling the basal area diameter distribution in stands of *Pinus silvestris* and *Picea abies*. Scand. J. For. Res. 10: 284-295.

Marschall, J., 1975: Hilfstafeln für die Forsteinrichtung. Österr. Agrarverlag, Vienna: 199 p.

Martin, G.L. and A.R. Ek, 1984. A comparison of competition measures and growth models for predicting plantation red pine diameter and height growth. Forest Science 30(3): 731-743.

Mayer, D.G. and Butler, D.G., 1993: Statistical validation. Ecol. Modelling, 68: 21-32.

McCullagh, P., Nelder, J.A., 1985: Generalized linear models. Chapman and Hall, London, New York.

Meyer, W.H., 1930: Diameter distribution series in even-aged forest stands. School of Forestry, Yale Univ. Bull. 28: 105 p.

Michaelis, K.A., 1910: Wie bringt Durchforsten die größere Stärke- und Wertzunahme des Holzes? 2.Auflage. Neudamm: Neumann.

Michailoff, I., 1943: Zahlenmäßiges Verfahren für die Ausführung der Bestandeshöhenkurven. Forstw. Cbl. u. Thar. Jahrb. 6: 273-279.

Mikulka, B. 1955: Versuche zur zahlenmäßigen Erfassung der Qualität von Waldbeständen. Mitt. der Schw. Anst. f. d. forstl. Vers. Wes. 23. Bd. 1.

Mitscherlich, G., 1971: Wald, Wachstum und Umwelt - Waldklima und Wasserhaushalt. J.D. Sauerländer's Verlag.

Moser, J. W., 1974: A system of equations for the components of forest growth. In: J. Fries (Hrsg): Growth models for tree and stand simulation, Royal College of Forestry. Research Notes Nr. 30: 56-76.

Murray, D. M., and Gadow, K. v., 1991: Relationship between the diameter distribution before and after thinning, For. Sci., 37(2), 552-559.

Murray, D.M. and Gadow, K.v. 1993: A flexible yield model for regional timber forecasting. Southern Journal of Applied Forestry 17 (2): 112-115.

Nagashima, I. and Kawata, N., 1994: A stem taper model including butt swell. J. Jpn. For. Soc. 76(4): 291-297.

Nagel, J., 1991: Einheitshöhenkurvenmodell für Roteiche, Allg. Forst- u. J.-Ztg., 162(1):16-18.

Nagel, J., 1994: Ein Einzelbaumwachstumsmodell für Roteichenbestände. Forst und Holz. 49. Jahrgang. Nr. 3, S. 69-75.

Nagel, J., and Biging, G.S., 1995: Schätzung der Parameter der Weibullfunktion zur Generierung von Durchmesserverteilungen, Allg. Forst- u. J.-Ztg., 166(9/10): 185-189.

Nagumo, H., Shiraishi, N. and Tanaka, M. 1981: Computer programming for the construction of a Sugi local yield table - Program Tusycs. Proc. XVII IUFRO World Congress, Kyoto, p. 103-114.

Nance, W., Grissom, J.E. and Smith, W.R. 1988: A new competition index based on weighted and constrained area potentially available. In: Ek, A.R., Shifley, S.R. & Burke, T.E. (eds): Forest growth modelling and prediction. USDA For. Serv. Gen. Techn. Rep. NC-120: 134-142.

Newham, R. M., 1966: Stand structure and diameter growth of individual trees in a young red pine stand. Can. Dept. for Rural Develop. Res. Notes 22: 4-5.

Niedersächsische Forstliche Versuchsanstalt, 1996: Stand der Versuchsflächen vom 1. Januar 1996.

O'Connor, A.J., 1935: Forest research with special reference to planting distances and thinning. Govt. Printer, Pretoria.

Opie, J.E., 1968: Predictability of individual tree growth using definitions of competing basal area. For. Sci. 14: 314-323.

Pardé, J., 1961: Dendrométrie. Editions de l'École Nationale des Eaux et des Forêts, Nancy.

Penttinen, A., Stoyan, D. and Henttonen, H.M., 1994: Marked point processes in forest statistics. For. Sci. 38 (4): 806-824.

Pielou, E.C., 1961: Segregation and symmetry in two-species populations as studied by nearest neighbour relations. J.Ecol.49: 255-269.

Pienaar, L. V., Page, H. and Rheney, J. W., 1990: Yield prediction for mechanically site-prepared slash pine plantations. Southern J. of Applied Forestry, 14(3):104-109.

Pienaar, L. V. and Harrison, W. M., 1988: A stand table projection approach to yield prediction in unthinned even-aged stands. For. Sci., 34(3):804-808.

Pienaar, L. V. andTurnbull, K. J., 1973: The Chapman-Richards generalization of von Bertalanffy's growth model for basal area growth and yield in even-aged stands, For. Sci.,19:2-22.

Pommerening, A., 1997: Eine Analyse neuer Ansätze zur Bestandesinventur in strukturreichen Wäldern. Diss. Fakult. für Forstwissenschaften und Waldökologie, Göttingen. Cuvillier Verlag, Göttingen. 187 S.

Pommerening, A., Gadow, K. v. and Lewandowski, A., 1997: A new approach to describing forest structures. For. Ecol. & Mgmt (submitted).

Pretzsch, H, 1992: Konzeption und Konstruktion von Wuchsmodellen für Rein- und Mischbestände. Schriftenreihe d. Forstw. Fak. . Univ. München, No. 115: 332 p.

Pretzsch, H., 1994: Analyse und Reproduktion räumlicher Bestandesstrukturen - Versuche mit dem Strukturgenerator STRUGEN. Schriften aus der Forstl. Fak. d. Univ. Göttingen u. d. Nds. Forstl. Vers. Anst., Band 114, J.D. Sauerländer's Verlag.

Preußner, K., 1974: Aufstellung und experimentelle Überprüfung mathematischer Modelle für die Entwicklung der Durchmesserverteilung von Fichtenbeständen. Diss. Tharandt.

Prodan, M., 1953: Verteilung des Vorrats gleichaltriger Hochwaldbestände auf Durchmesserstufen. Allg. Forst u. Jagdzeitung 129: 15-33.

Prodan, M., 1965: Holzmeßlehre. J.D. Sauerländer's Verlag, Frankfurt a.M.

Pukkala, T., 1989: Prediction of tree diameter and height in a Scots pine stand as a function of the spatial pattern of trees. Silva Fenn. 23: 83-99.

Pukkala, T. and Kolström, T., 1988: Simulation of the Development of Norway Spruce Stands using a Transition Matrix. Forest Ecology and Management 25: 255-267.

Puumalainen, J., 1996: Die Beta-Funktion und ihre analytische Parameterbestimmung für die Darstellung von Durchmesserverteilungen. Arbeitspapier 15-96, Inst. f. Forsteinrichtung u. Ertragskunde d. Univ. Göttingen: 12 p.

Quicke, H.E., Meldahl, R.S. and Kush, J.S., 1994: Basal area growth of individual trees - a model derived from a regional longleaf pine growth study. For. Sci. 40 (3): 528-542.

Ramirez-Maldonado, H., Bailey, R.L. and Borders, B.B., 1988: Some implications of the algebraic difference form approach for developing growth models. In: Ek, A.R., Shifley, S.R. and Burke, T.E. (eds): Forest growth modelling and prediction. USDA For. Serv. Gen. Techn. Rep. NC-120: 731-738.

Raven, P.H., Evert, R.F., Eichhorn, S.E., 1987: Biology of plants. Worth Publishers Inc.

Reed, D.D. and Green, E.J., 1984: Compatible stem taper and volume ratio equations. For. Sci. 30 (4): 977-990.

Rennols, K and Smith, W.R., 1993: Zone of influence models for inter tree forest competition. In: Rennols, K. (ed.) 1993: Stochastic spatial models in forestry. Proc. of a IUFRO S4.11 Conf. held in Thessaloniki, Greece; published by The University of Greenwich: 27-36.

Riemer, Th., Gadow, K. v. and **Sloboda, B.**, 1995: Ein Modell zur Beschreibung von Baumschäften. Allg. Forst-und J.-Ztg. *166(7)*, 144-147.

Rodriguez Soalleiro, R., 1995: Crecimiento y producción de masas forestales regulares de *Pinus pinaster Ait.* en Galicia - alternativas selvicolas posibles. Escuela Tecnica Superior de Ingenieros de Montes: 297 p.

Rojo y Alboreca, A. and **Montero Gonzalez, G.**, 1996: El pino silvestre en la Sierra da Guadarrama. Ministerio de Agricultura, Pesca y Alimentación, Madrid: 293 p.

Römisch, K., 1983: Ein mathematisches Modell zur Simulation von Wachstum und Durchforstung gleichaltriger Reinbestände. Diss. Tharandt.

Rudra, A.B., 1968: A stochastic model for the prediction of diameter distributions of even-aged forest stands. OPSEARCH. Journal of the Operational Society of India. 5 (2): 59-73.

Saborowski, J., 1982: Entwicklung biometrischer Modelle zur Sortimentenprognose. Diss. Univ. Göttingen: 146 p.

Schädelin, W., 1942: Die Auslesedurchforstung als Erziehungsbetrieb höchster Wertleistung. 3rd edition, Bern.

Schober, R., 1952: Massentafeln zur Bestimmung des Holzgehaltes stehender Waldbäume und Waldbestände. Berlin.

Schober, R., 1957: Deutung und Aussage der Durchforstungsversuche, II. Die Buchendurchforstungsversuche. AFZ 12: 321-324, 389-394.

Schober, R., 1987: Ertragstafeln wichtiger Baumarten. J. D. Sauerländer's Verlag. Frankfurt am Main.

Schober, R., 1991: Eclaircies par le haut et arbres d'avenir. Rev. For. Fr. XLIII: 385-401.

Schübeler, D., Nagel, J., Pommerening, A. and **Gadow K. v.**, 1995: Modellierung des standortbezogenen Wachstums der Fichte. Unveröff. Manuskript, Inst. f. Forsteinr. u. Ertragsk., d. Univ. Göttingen: 68 p.

Schulz, H., 1954: Untersuchung über Bewertung und Gütemerkmale des Eichenholzes aus verschiedenen Wuchsgebieten. Schriftenreihe der Forstlichen Fakultät der Universität Göttingen. J. D. Sauerländer's Verlag, Frankfurt am Main.

Schumacher, F. X., 1939: A new growth curve and its application to timber-yield studies. J. For. 37:819-820.

Schütz, J.-Ph., 1989: Der Plenterbetrieb. Unterlage zur Vorlesung Waldbau III, ETH Zürich: 54 S.

Schwappach, A., 1905: Untersuchungen über die Zuwachsleistungen von Eichenhochwaldbeständen in Preußen. Verlag J. Neumann, Neudamm.

Schweizerisches Landesforstinventar, 1988: Berichte Nr. 1986: Ergebnisse der Aufnahme 1982-1986. Eidgenössische Anstalt für das forstliche Versuchswesen. Birmensdorf.

Seebach, C., 1846: Ertragsuntersuchungen im Buchenhochwalde. Krit. Blätter f. Forst- u. Jagdwiss. 23: 74-88.

Sharpe, P.J.H., 1990: Forest modeling approaches - compromises between generality and precision. In: Dixon, Mehldal, Ruark, Warren (eds.): Process Modeling of Forest Growth Responses to Environmental Stress; Timber Press, Portland, Oregon: 21-32.

Shiver, B.D., 1988: Sample sizes and estimation methods for the Weibull distribution for unthinned slash pine plantation diameter distributions. For. Sci. 34 (3): 809-814.

Shvidenko, A., Venevsky, S., Raille,G. and **Nilsson, S.**, 1995: A system for evaluation of growth and mortality in Russian forests. Water, Air and Soil Pollution 82: 333-348.

Shvets, V. and **Zeide, B.**, 1996: Investigating parameters of growth equations. Can. J. For. Res. 26 (11): 1980-1990.

Skovsgaard, J.P., 1997: Tyndingsfri drift af sitkagran. The Research Series, Forskningcentret for Skov & Landskab, Nr. 19: 525 pp.

Sloboda, B., 1971: Zur Darstellung von Wachstumsprozessen mit Hilfe von Differentialgleichungen erster Ordnung. Mitt. Bad. - Württemb. Forstl. Vers. u. Forsch. Anstalt. Heft 32.

Sloboda, B., 1976: Mathematische und stochastische Modelle zur Beschreibung der Statik und Dynamik von Bäumen und Beständen - insbesondere das bestandesspezifische Wachstum als stochastischer Prozeß. Habil. Schrift. Univ. Freiburg.

Sloboda, B., 1984: Bestandesindividuelles biometrisches Schaftformmodell zur Darstellung und zum Vergleich von Formigkeit und Sortimentausbeute sowie Inventur. Tagungsbericht d. Sektion Ertragskunde, Neustadt.

Sloboda, B. and **Pfreundt, J.**, 1989: Baum- und Bestandeswachstum: Ein systemanalytischer räumlicher Ansatz mit Versuchsplanungskonsequenzen für die Durchforstung und Einzelbaumentwicklung. Deutscher Verband Forstlicher Forschungsanstalten Sektion Ertragungskunde. Attendorn/Olpe.

Sloboda, B., Gaffrey, D. and **Matsumura, N.**, 1993: Regionale und lokale Systeme von Höhenkurven für gleichaltrige Waldbestände. Allg.Forst- u. J.-Ztg., 164(12):225-228.

Smalley, G.W. and **Bailey, R.L.**, 1974: Yield tables and stand structure for shortleaf pine plantations in the Tennessee, Alabama and Georgia highlands. USDA For. Service Res. Paper SO-97.

Smaltschinski, T., 1981: Bestandesdichte und Verteilungsstruktur. PhD Diss. Univ. of Freiburg.

Smaltschinski, T., 1997: Großregionale Holzaufkommensprognosen und nationale Forstinventuren. Unpublished manuscript, University of Göttingen: 199 p.

Soares, P., Tomé, M., Skovsgaard, J.P. and Vanclay, J.K. 1995: Evaluating a growth model for forest management using continuous forest inventory data. For. Ecol. and Mgmt. 71: 251-265.

Späth, H., 1973: Spline-Algorithmen zur Konstruktion glatter Kurven und Flächen. Oldenbourg, München.

Speidel, G., 1955: Die Wertklasse als Gütemaßstab in der Forsteinrichtung. FA, 217-224.

Spiecker, H., 1989: Eichenwertholzerzeugung - zur Steuerung des Dickenwachstums und der Astreinigung von Trauben- und Stieleichen. Habil. thesis University of Freiburg/Germany.

Spiecker, M., 1994: Wachstum und Erziehung wertvoller Waldkirschen. Mitt. d. Forstl. Vers. u. Forschungsanstalt Baden-Württemberg, No 181.

Staebler, G.R., 1951: Growth and spacing in an even-aged stand of Douglas-fir. Master's thesis, Univ. of Michigan.

Stage, A. R., 1963: A mathematical approach to polymorphic site index curves for Grand-Fir. For. Sci., 9(2):167-180.

Steingaß, F., 1995: Beschreibung der Schaftprofile von Douglasien. Diplomarbeit, Forstw. Fachbereich, Univ. Göttingen.

Sterba, H., 1975: Assmanns Theorie der Grundflächenhaltung und die "Competition-Density-Rule" der Japaner Kira, Ando und Tadaki. Centralblatt für das gesamte Forstwesen. 92(1):46-62.

Sterba, H., 1987: Estimating potential density from thinning experiments and inventory data. For. Sci. 33 (4): 1022-1034.

Sterba, H. 1991: Forstliche Ertragslehre. Lecture notes (Heft 4): 159 p.

Sterba, H. and Monserud, R.A., 1997: Applicability of the forest stand growth simulator Prognaus for the Austrian part of the Bohemian Massif. Ecol. Mod. 98: 23-34.

Suzuki, T., 1971: Forest transition as a stochastic process. Mitt. FBVA Wien. Heft 91: 137-150.

Tewari, V. P. and Gadow, K. v., 1995: Fitting a bivariate distribution to diameter-height data. Working paper 6/95. Institut Forsteinrichtung und Ertragskunde, Univ. Göttingen.

Tome, M. and Burkhart, H.E., 1989: Distance-dependent competition measures for predicting growth of individual trees. Forest Science 35(3): 816-831.

Tompo, E., 1986: Models and methods for analysing spatial patterns of trees. Communicationes Instituti Forestalis Fenniae No. 138, Helsinki: 65pp.

Trepl, L., 1994: Competition and coexistence - on the historical background in ecology and the influence of economy and social sciences. Ecol. Modelling 75/76: 99-110.

Trincado, G. and Gadow, K. v., 1996: Zur Sortimentschätzung stehender Laubbäume. Zentralblatt für das gesamte Forstw. 113 (1): 27-38.

Trincado, G., 1996: Modellierung der Schaftform von Fichten (*Picea abies*) und Buchen (*Fagus sylvatica*). Wissenschaftliche Arbeit zur Erlangung des Grades Magister der tropischen Forstwirtschaft der Georg-August-Universität, Göttingen. 50 p.

Upton, G. and Fingleton, B., 1990: Spatial data analysis by example. Wiley.

Vanclay, J.K., 1994: Modelling forest growth - applications to mixed tropical forests. CAB International, Wallingford, UK.

Van Laar, A. and Akça, A., 1997: Forest Mensuration. Cuvillier, Göttingen: 418 pp.

Villarino Urtiaga, J.J. and Riesco Muños, G., 1997: La relacion diametro de copa - diametero normal en *Betula celtiberica* Rothm. et Vas. Proc. I Congreso Forestal Luso.

Wenk, G., Antanaitis, V. and Smelko, S., 1990: Waldertragslehre. Deutscher Landwirtschaftsverlag. Berlin.

Wensel, L.C. and Koehler, J.R., 1985: A tree growth projection system for Northern California coniferous forests. Northern California Forest Yield Cooperative Res. Note No. 12: 30p.

Wiant, H.V., Wood, G.B. and Gregoire, T.G., 1992: Practical guide for estimating the volume of a standing sample tree using either importance or centroid sampling. For. Ecol. and Mgmt. 49: 333-339.

Wiedemann, E., 1935: Über die Vereinfachung der Höhenermittlung bei der Vorratsaufnahme. Mitt. a. Forstw. u. Forstwiss.: 387-412.

Wiegard, C., Netzker, D. and Gadow, K. v., 1997: Die Erdstückmethode der Wertinventur. Forstarchiv 68: 144-148.

Wimmenauer, 1902: Die diesjährige Versammlung des Vereins deutscher forstlicher Versuchsanstalten. Beilage: Anleitung zur Ausführung von Durchforstungs- und Lichtungsversuchen. AFJZ 422-5.

Wu, H., Sharpe, P., Walker, J. and Penridge, L., 1985: Ecological field theory - a spatial analysis of resource interference among plants. Ecol. Modelling 29: 215-243.

Wykoff, W. R., Crookston, N. L. and Stage, A. R., 1982: User's guide to the stand prognosis model. USDA For. Serv., Gen. Tech. Rep. INT-133.

Yu, X. T., 1982: *Cunninghamia lanceolata*. Science and Technology, Fujian (in Chinese).

Zeide, B., 1993: Analysis of growth equations. For. Sci. 39 (3): 594-616.

Zimmermann, H.-J., 1991: Fuzzy set theory and its applications. Kluwer Academic Publishers. 2. Auflage.

Zucchini, W. and Gadow, K. v., 1995: Two indices of agreement among foresters selecting trees for thinning. Forest and Landscape Research 1: 199-206.

Index

FORESTRY SCIENCES

1. P. Baas (ed.): *New Perspectives in Wood Anatomy.* Published on the Occasion of the 50th Anniversary of the International Association of Wood Anatomists. 1982
    ISBN 90-247-2526-7
2. C.F.L. Prins (ed.): *Production, Marketing and Use of Finger-Jointed Sawnwood.* Proceedings of an International Seminar Organized by the Timber Committee of the UNECE (Halmar, Norway, 1980). 1982
    ISBN 90-247-2569-0
3. R.A.A. Oldeman (ed.): *Tropical Hardwood Utilization.* Practice and Prospects. 1982
    ISBN 90-247-2581-X
4. P. den Ouden (in collaboration with B.K. Boom): *Manual of Cultivated Conifers.* Hardy in the Cold- and Warm-Temperate Zone. 3rd ed., 1982
    ISBN Hb 90-247-2148-2; Pb 90-247-2644-1
5. J.M. Bonga and D.J. Durzan (eds.): *Tissue Culture in Forestry.* 1982   ISBN 90-247-2660-3
6. T. Satoo: *Forest Biomass.* Rev. ed. by H.A.I. Madgwick. 1982   ISBN 90-247-2710-3
7. Tran Van Nao (ed.): *Forest Fire Prevention and Control.* Proceedings of an International Seminar Organized by the Timber Committee of the UNECE (Warsaw, Poland, 1981). 1982
    ISBN 90-247-3050-3
8. J.J. Douglas: *A Re-Appraisal of Forestry Development in Developing Countries.* 1983
    ISBN 90-247-2830-4
9. J.C. Gordon and C.T. Wheeler (eds.): *Biological Nitrogen Fixation in Forest Ecosystems.* Foundations and Applications. 1983   ISBN 90-247-2849-5
10. M. Németh: *Virus, Mycoplasma and Rickettsia Diseases of Fruit Trees.* Rev. (English) ed., 1986
    ISBN 90-247-2868-1
11. M.L. Duryea and T.D. Landis (eds.): *Forest Nursery Manual.* Production of Bareroot Seedlings. 1984; 2nd printing 1987   ISBN Hb 90-247-2913-0; Pb 90-247-2914-9
12. F.C. Hummel: *Forest Policy.* A Contribution to Resource Development. 1984
    ISBN 90-247-2883-5
13. P.D. Manion (ed.): *Scleroderris Canker of Conifers.* Proceedings of an International Symposium on Scleroderris Canker of Conifers (Syracuse, USA, 1983). 1984 ISBN 90-247-2912-2
14. M.L. Duryea and G.N. Brown (eds.): *Seedling Physiology and Reforestation Success.* Proceedings of the Physiology Working Group, Technical Session, Society of American Foresters National Convention (Portland, Oregon, USA, 1983). 1984   ISBN 90-247-2949-1
15. K.A.G. Staaf and N.A. Wiksten (eds.): *Tree Harvesting Techniques.* 1984
    ISBN 90-247-2994-7
16. J.D. Boyd: *Biophysical Control of Microfibril Orientation in Plant Cell Walls.* Aquatic and Terrestrial Plants Including Trees. 1985   ISBN 90-247-3101-1
17. W.P.K. Findlay (ed.): *Preservation of Timber in the Tropics.* 1985   ISBN 90-247-3112-7
18. I. Samset: *Winch and Cable Systems.* 1985   ISBN 90-247-3205-0
19. R.A. Leary: *Interaction Theory in Forest Ecology and Management.* 1985
    ISBN 90-247-3220-4
20. S.P. Gessel (ed.): *Forest Site and Productivity.* 1986   ISBN 90-247-3284-0
21. T.C. Hennessey, P.M. Dougherty, S.V. Kossuth and J.D. Johnson (eds.): *Stress Physiology and Forest Productivity.* Proceedings of the Physiology Working Group, Technical Session, Society of American Foresters National Convention (Fort Collins, Colorado, USA, 1985). 1986
    ISBN 90-247-3359-6

FORESTRY SCIENCES

FORESTRY SCIENCES

KLUWER ACADEMIC PUBLISHERS – DORDRECHT / BOSTON / LONDON